SOLAR DISTILLATION

Other Pergamon Titles of Interest

ARIZONA STATE UNIVERSITY LIBRARY	Solar Energy Index
DE WINTER	Sun: Mankind's Future Source of Energy
EGGERS-LURA	Solar Energy For Domestic Heating and Cooling
EGGERS-LURA	Solar Energy in Developing Countries
FAZZOLARE & SMITH	Beyond The Energy Crisis: Opportunity and Challenge (ICEUM 3) - 4 Volumes
GOODMAN & LOVE	Small Hydroelectric Projects For Rural Development
HALL	Solar World Forum - 3 Volumes
HOWELL	Your Solar Energy Home
JAGER	Solar Energy Applications in Houses
MARTINEZ	Solar Cooling and Dehumidifying
McVEIGH	Sun Power, 2nd Ed.
PALZ & STEEMERS	Solar Houses in Europe
SMITH	Energy Management Principles
STAMBOLIS	Solar Energy in the 80s
TAHER	Energy: A Global Outlook
TWIDELL	Energy for Rural and Island Communities

Related Journals - *Free Specimen Copies Gladly Sent on Request*
ENERGY
ENERGY CONVERSION AND MANAGEMENT
INTERNATIONAL JOURNAL OF HYDROGEN ENERGY
PROGRESS IN WATER TECHNOLOGY
SOLAR ENERGY
SPACE SOLAR POWER REVIEW
SUN AT WORK IN BRITAIN
SUN WORLD
WATER RESEARCH
WATER SUPPLY AND MANAGEMENT

SOLAR DISTILLATION

A Practical Study of a Wide Range of Stills and Their Optimum Design, Construction and Performance

by

M. A. S. MALIK
Renewable Energy Specialist, World Bank, USA

G. N. TIWARI
Centre of Energy Studies, Indian Institute of Technology

A. KUMAR
Centre of Energy Studies, Indian Institute of Technology

M. S. SODHA
Professor and Deputy Director, Indian Institute of Technology, New Delhi, India

PERGAMON PRESS

OXFORD · NEW YORK · TORONTO · SYDNEY · PARIS · FRANKFURT

U.K.	Pergamon Press Ltd., Headington Hill Hall, Oxford OX3 0BW, England
U.S.A.	Pergamon Press Inc., Maxwell House, Fairview Park, Elmsford, New York 10523, U.S.A.
CANADA	Pergamon Press Canada Ltd., Suite 104, 150 Consumers Rd., Willowdale, Ontario M2J 1P9, Canada
AUSTRALIA	Pergamon Press (Aust.) Pty. Ltd., P.O. Box 544, Potts Point, N.S.W. 2011, Australia
FRANCE	Pergamon Press SARL, 24 rue des Ecoles, 75240 Paris, Cedex 05, France
FEDERAL REPUBLIC OF GERMANY	Pergamon Press GmbH, 6242 Kronberg-Taunus, Hammerweg 6, Federal Republic of Germany

Copyright © 1982 Pergamon Press Ltd.

All Rights Reserved. No part of this publication may be reproduced, stored in a retrieval system or transmitted in any form or by any means: electronic, electrostatic, magnetic tape, mechanical, photocopying, recording or otherwise, without permission in writing from the publishers.

First edition 1982

Library of Congress Cataloging in Publication Data
Solar distillation.
Bibliography: p.
1. Solar stills. I. Malik, M. A. S.
TP159.D5S65   1982   660.2'8425   81-17767
AACR2

British Library Cataloguing in Publication Data
Solar distillation
1. Solar stills 2. Saline water conversion
—Distillation process
I. Malik, M. A. S.
628.1'672   TD480
ISBN 0-08-028679-8

In order to make this volume available as economically and as rapidly as possible the authors' typescripts have been reproduced in their original forms. This method unfortunately has its typographical limitations but it is hoped that they in no way distract the reader.

Printed in Great Britain by A. Wheaton & Co. Ltd., Exeter

PREFACE

The vital importance of supplying drinking water to the whole human population has been recognized by the U.N.O. which has declared 1981-90 as the International Drinking Water Supply and Sanitation Decade (IDWSSD). At the end of this program it is contemplated that the drinking water requirements of at least half the world's population will have been met. Considering the fact that an estimated 2 billion people go without fresh water today, this is indeed an ambitious goal. Consumption of water, unfit for drinking on account of bacteria/germs or excess of salts, is a major health hazard, which alone is a sufficiently strong reason to undertake a massive scheme to supply fresh water to all, who do not have it.

The system for supply of drinking water in a region with abundant fresh water (lakes/rivers/underground etc.,) is very different from one corresponding to places without any such source e.g. deserts, marshy lands etc., which may have only brackish water. Economic considerations may render distillation of saline/brackish water a better choice than supplying water by trucks or by laying long pipe-lines from a distant source. It may happen that, with a modest supply of fresh water, a region, hitherto considered as waste-land, becomes economically exploitable. There are several methods of distillation to choose from. Most of the conventional water distillation plants are energy-intensive and require scarce electric power or fossil fuel for operation; however solar energy despite being a much lower grade energy, is ideally suited for this job. The choice of solar energy has become even more attractive on account of the ten-fold rise in petroleum crude prices as compared to that in the beginning of 1970. Further, solar energy is not a monopoly and the technology involved in distillation of saline water using solar energy is relatively simple; a very large component of the fabrication (men and materials) of solar stills can be indigenous and maintenance can be carried out by semiskilled or unskilled operators.

Considerable amount of literature, dating back to the Arab alchemists of 1551, exists on solar stills; the work has attracted renewed interest in recent times. Unfortunately, no exhaustive study has been published till now, despite the importance of the topic. Although some review articles have been published, they are either outdated or are inadequate in scope, giving a glimpse of only few aspects of the subject. The need for a relatively unified and comprehensive account of the various aspects of solar distillation has been felt by the solar-energy community for a long time; the present monograph is an attempt to this end.

A brief historical review has been presented in the introductory chapter along with a qualitative discussion on the fundamental aspects of solar stills and the identification of basic energy transfer mechanisms. Chapter 2 consists of a discussion on the basic heat and mass transfer relations, necessary for a proper understanding of the operation of a still. The single basin solar still has been studied in detail in Chapter 3, using the steady state theory as well as the periodic heat transfer model developed in Chapter 2; transient analysis has also been given. Besides these, the effect of using various dyes on the performance of the still, and the nocturnal production have also been discussed. Other aspects such as materials and effect of various meteorological and still parameters on the performance of the still have also been considered. Chapter 4 discusses the multi-effect solar stills and the double basin solar still in detail. Inclining a solar still with respect to the horizontal, as is done with flat-plate solar collectors, increases the distillate output of a still; this aspect has been examined in Chapter 5. In particular the "Multiple-Wick Solar Still" has been discussed in some detail as it seems to offer significant advantages over the other designs. Chapter 6 discusses various other stills of rather exotic designs. The solar still green-house combination, which can lead to selective agriculture in regions having a supply of brackish water has been discussed in Chapter 7, a periodic analysis of a "Still on the roof" concept is the quantitative highlight of the chapter. The monograph has been rounded-off with a discussion of the economic aspects of using solar stills.

M.K.S. system of units has been adopted throughout the book. The cost prices have been given in U.S. Dollars (wherever possible). A list of symbols used in the text has been compiled in the end in conformity with the usual practice in the solar energy community.

The IITD group of co-authors is grateful to Professor O. P. Jain, Director, for his kind encouragement. Thanks are also due to Dr. S. D. Gomkale for writing Appendix A.

The helpful editing by Dr. G. Umesh is gratefully acknowledged.

New Delhi.

MASM
GNT
AK
MSS

CONTENTS

1.	INTRODUCTION	1
	1.1 Importance	1
	1.2 Historical Review	2
	1.3 Basic Principles	5
2.	BASIC HEAT AND MASS TRANSFER RELATIONS	8
	2.1 Internal Transfer	9
	2.1.1 Convection	9
	2.1.2 Evaporation	11
	2.1.3 Radiation	13
	2.2 External Transfer Mode	14
	2.2.1 Convection and Radiation	14
	2.2.2 Periodic Heat Transfer in Conducting Media	15
	2.2.3 Periodic Boundary Conditions	16
3.	SINGLE BASIN SOLAR STILL	18
	3.1 Background	18
	3.2 Solar Radiation Balance	20
	3.3 Analysis of an Ideal Still	22
	3.4 Periodic Theory of Solar Still	23
	3.4.1 (a) Ground Still	23
	3.4.1 (b) Mounted Still	28
	3.4.2 Comparison of Experiment and Theory	28
	3.5 Effect of Dye on the Performance of a Solar Still	33
	3.5.1 Numerical Calculations and Discussion	33
	3.6 Transient Analysis of Solar Still	37
	3.6.1 Background	37
	3.6.2 Analysis	37

Contents

3.7	Nocturnal Production of Solar Stills	39
	3.7.1 Tubular Solar Still	40
	3.7.2 Single Basin Solar Still	40
	a. A Simplified Mathematical Model	42
	b. Experimental Investigation	44
	c. Rigorous Analytical Model	46
	d. Numerical Results and Discussion	49
3.3	Parametric Studies	50
	3.8.1 Experience with Plastic Covers	50
	3.8.2 Other Materials	53
	3.8.3 Effect of Meteorological and Still Parameters	54
	a. Effect of wind velocity	54
	b. Effect of ambiant temperature	54
	c. Effect of solar radiation and loss coefficient	54
	d. Effect of double glass cover and cover inclination	56
	e. Effect of thermal capacity on output	56
	f. Effect of salt concentration on output	56
	g. Effect of charcoal pieces on the performance of still	57
	h. Effect of the formation of algae and mineral layers on water and basin liner surface on productivity of still	58
4.	**MULTIPLE EFFECT SOLAR STILL**	**59**
	4.1 General Consideration	59
	4.2 Diffusion Still	60
	4.3 Chimney Type Solar Still	61
	4.4 Heated Head Solar Still	61
	4.5 A Simple Multi-effect Basin Type Solar Still	62
	4.5.1 Design Principle	62
	4.5.2 Construction	62
	4.5.3 Performance	63
	4.6 Three-Effect Multiple Solar Still	64
	4.6.1 Construction	64
	4.6.2 Results	65
	4.6.3 Discussion	66
	4.7 Double Basin Solar Still	67
	4.7.1 Background	67
	4.7.2 Construction	68
	4.7.3 Analysis	68
	4.7.4 Comparison with Experiment	70

Contents ix

5. **INCLINED SOLAR STILL** — 72
   - 5.1 Tilted Tray or Inclined-stepped Solar Still — 72
   - 5.2 Stepped Inclined Still — 75
   - 5.3 Multi Basin System — 76
     - 5.3.1 Basin type solar still — 76
   - 5.4 Tilted Single Wick Stills — 79
   - 5.5 Multiple-ledge Tilted Stills — 80
   - 5.6 Wick Type Collector-Evaporator Still — 81
   - 5.7 Simple Multiple Wick Solar Still — 81
     - 5.7.1 Experiment — 82
     - 5.7.2 Analysis — 85
     - 5.7.3 Numerical Results and Discussions — 87
     - 5.7.4 Advantages of Multiple Wick Still — 88

6. **OTHER DESIGNS OF SOLAR STILLS** — 90
   - 6.1 Life Raft Type Solar Still — 90
   - 6.2 Film Covered Solar Still — 91
   - 6.3 Wiping Spherical Still — 92
   - 6.4 Concentric Tube Solar Still — 93
   - 6.5 Solar Earth Water Stills — 94
   - 6.6 Combined Solar Collector - Basin Type Still Systems — 96
   - 6.7 Air-supported Plastic Still — 97
   - 6.8 Effect of Floating Mat on Productivity of a Still — 98
   - 6.9 Solar Still with Reflector — 98
   - 6.10 Extruded Plastic Still — 99

7. **SOLAR STILL GREEN HOUSE COMBINATION** — 100
   - 7.1 Solar Still Green House Combination: Texas Experience — 100
   - 7.2 Materials for Green House Solar Still Systems — 101
     - 7.2.1 Measurement of Photosynthetic Activity — 102
     - 7.2.2 Estimation of Still Production — 104
   - 2.3 Still on Roof: Analysis — 104
     - 7.3.1 Background — 104
     - 7.3.2 Analysis — 106
     - 7.3.3 Numerical Calculations and Discussion — 109

8. **ECONOMIC ASPECTS OF SOLAR DISTILLATION** — 110
   - 8.1 Cost of Product Water — 111
   - 8.2 Early Economic Analysis — 112
   - 8.3 Russian Experience — 114
   - 8.4 Indian Experience — 115
   - 8.5 Solar Distillation versus Conventional Processes — 118

Contents

9. RECOMMENDATIONS FOR FUTURE RESEARCH   121
10. APPENDIXES   122
11. NOMENCLATURE   127
12. REFERENCES   131
13. SELECTED BIBLIOGRAPHY   136
14. AUTHOR INDEX   171
15. SUBJECT INDEX   173

1. INTRODUCTION

1.1. IMPORTANCE

Water is a basic necessity of man along with food and air; the importance of supplying hygienic potable/fresh* water can hardly be overstressed. Man has been dependent on rivers, lakes and underground water reservoirs for fresh water requirements in domestic life, agriculture and industry. However, use of water from such sources is not always possible or desirable on account of the presence of large amount of salts and harmful organisms. The impact of many diseases afflicting mankind can be drastically reduced if fresh hygienic water is provided for drinking. Further, the rapid industrial growth and population explosion all over the world has resulted in a large escalation of demand for fresh water; this invariably leads to acute fresh water shortages since the natural sources of water can meet the demands to a very limited extent. Added to this is the problem of pollution of the rivers and lakes by the industrial wastes and the large amounts of sewage. Thus there is scarcity of fresh water even in cities, towns and villages near lakes and rivers. Dangerous pollutants left on open ground also find their way into the underground reservoirs along with rainwater. In fact on a global scale man-made pollution of natural sources of water is turning out to be the single largest cause for the fresh water shortage. Besides this there are several regions on the earth, e.g. the deserts, which have inhospitable climatic conditions and have only brackish water sources. In such places fresh water will have to be provided not only for domestic use, but also for agricultural needs. It would be no exaggeration to say that by the end of this century, supply of adequate quantities of fresh potable water could become one of the most serious problems confronting man.

With the official launching in New York, U.S.A. on 10th November, 1980, of the International Drinking Water Supply and Sanitation Decade (IDWSSD), the United Nations Organisation set into motion a major initiative that should have a direct impact on at least half the world's population by 1990. According to present estimates, over 2000 million people are without reasonable access to a safe and adequate water supply. Developing countries (e.g. India) have

---

*Less than or about 500 parts per million of salt.

given utmost priority to rural water supply in their development plans. Major U.N. organisations like UNDP, WHO and the World Bank are actively involved in promoting projects aimed at supplying drinking water in Indian villages. Between 25 to 30 per cent of the UNICEF's assistance to programs for children is invested in water supply. The UNDP and World Bank have been involved in a global project aimed at low cost water supply techniques and to facilitate the testing and selection of water supply hand pumps.

The only inexhaustible sources of water are the oceans. Besides, there are also several, hitherto unused, big lakes, inland seas and underground natural reservoirs containing salt/brackish water. The chief drawback, obviously, is the very large salinity of such water. One of the attractive schemes to tackle the problem of water shortage is the distillation of such water resulting in desalination; this water may be mixed with brackish water (if hygienically desirable) to increase the amount of fresh water and bring the concentration of salts to around 500 parts per million. The conventional distillation processes such as multi-effect evaporation, multi-stage fresh evaporation, thin film distillation, reverse osmosis and electrodialysis are not only energy intensive but are also uneconomical for not too large demands of fresh water. However, the developments in the use of solar energy have demonstrated that it is ideally suited for desalination, when the demand of fresh water is not too large. The rapid escalation in the costs of fuels has made the solar alternative more attractive; in certain remote arid regions, this may be the only alternative. The least that can be said in favor of solar distillation (distillation of saline water by the use of solar energy) is that it is a viable option for providing hygienic potable water for a single house or a small community in most places of the earth. Further, the development of green-houses has resulted in minimizing the water requirements for carrying on agriculture on a small scale; thus solar desalination can support small scale agriculture in regions having only brackish water sources.

1.2. HISTORICAL REVIEW

Solar distillation has been in practise for a long time. The earliest documented work is that of the Arab alchemists in 1551 (Mouchot, 1869). In his review, Mouchot writes:*

"One uses glass vessels for the solar distillation operation . . . . According to the Arab alchemists, polished Damascus concave mirrors should be used for solar distillation."

In their historical review on desalination of water, Nebbia and Menozzi (1966) mention the work of Della Porta which he published in 1589. The apparatus used by Della Porta is illustrated in Fig. 1.1 and his own description of the experiment is as follows:

---

*Translated from original text in French.

Introduction

"About Distilling*

.... insert these into wide earthen pots full of water, so that the vapors may thicken more quickly into water. Turn all this apparatus, when it has been very carefully prepared, to the most intense heat of the sun's rays: for immediately they dissolve into vapors, and will fall drop by drop into the vases which have been placed underneath. In the evening, after sunset, remove them and fill with new herbs. Knot-grass, also commonly called "sparrow's tongue", when it has been cut up and distilled is very good for inflammation of the eyes and other afflictions. From the ground-pine is produced a liquid which will end all convulsions if the sick man washes his limbs with it: and there are other examples too numerous to mention. The picture demonstrates the method of distilling."

Fig. 1.1. Solar distillation apparatus of Della Porta.
(After Della Porta, 1589).

The great French chemist Lavoisier (1862) used large glass lenses, mounted on elaborate supporting structures, to concentrate solar energy on the contents of distillation flasks. The use of silver or aluminium coated glass reflectors to concentrate solar energy for distillation purposes has also been described by Mauchot (1869). Thus, it appears that in 19th century scientists were familiar with harnessing solar energy for distillation not only by direct exposure to the sun but also by concentrating sun's rays by means of mirrors and lenses.

The conventional solar distillation apparatus, (commonly known as the solar still), was first designed and fabricated in 1872 near Las Salinas in Northern Chile by Carlos Wilson, a Swedish engineer. This was a large basin-type solar still, meant for supplying fresh water to a nitrate mining community (Harding, 1883). Several wooden bays of size 1.14m × 61.0m were joined together to yield a total surface of area 4700 m^2, which was covered with glass. The bottom of the bays, exposed to the sun, was blackened with logwood dye and alum. Brackish water was poured into the bays which, upon evaporation, aided

---

*English translation from Latin text.

by solar energy, condensed over the glass cover and trickled down into the collectors. This device was in operation for about 40 years and yielded more than 4.9 Kg of distilled water per square meter of the still surface on a typical summer day (Harding, 1883). It is worth noting that this output compares very well with the distilled water output from the present day solar stills. The device was, however, abandoned later on when cheaper and more convenient methods of desalination of water came into vogue. The major problem with the still was the rapid accumulation of salts in the basin making regular flushing of the still a necessity.

No work on solar distillation seems to have been published after 1880's till the end of the First World War. With the renewal of interest, several types of devices have been described, e.g. roof type, V-covered, tilted wick, inclined tray, suspended envelope, tubular and air inflated stills etc. Use of metal coated reflectors as solar concentrators for application in solar distillation has been described by Kausch (1920); Pasteur (1928) also used several concentrators to focus solar rays onto a copper boiler containing water. The steam generated from the boiler was piped to a conventional water-cooled condenser in which distilled water was accumulated. The efficiency of this device was less than 50%. Abbot (1938) used cylindrical parabolic reflectors (aluminium coated surface) to focus solar energy onto evacuated tubes containing water. He also devised a 'clock-work' arrangement to track the motion of the sun. Although efficiencies as high as 80% could be achieved, the boiling of water in the tubes created some problems. During the Second World War, Maria Telkes (1945) developed air inflated plastic stills for the U.S. Navy and Air Force for use in the emergency life-rafts. The arrangement consisted of an inflatable transparent plastic bag inside which a porous felt pad was suspended with collector bottles placed at the bottom. Whenever it was required to be used, the felt pad was to be saturated with sea water. Water evaporates from the pad on account of incident insolation and condenses on the interior surface of the bag. Distilled water gets collected in the bottles kept at the bottom. As many as 200,000 of these stills were used by the Navy during the war.

The next stage was to improve the operating efficiencies of the various types of solar distillation devices. Forced air circulation was tried in stills to enhance the vapor condensation rate. Several investigators have attempted to make use of the latent heat of vaporization in either multiple-effect systems or for preheating the brine to increase the output of the stills. Several large scale distillation plants and integrated schemes for combining electric power generation and desalination of water have also been suggested as a way of improving the overall operating efficiency of the plant.

The basin-type solar still (also called as the greenhouse type, roof-type, simple or conventional type) is in the most advanced stage of development. Several workers have investigated the effect of climatic, operational and design parameters on the performance of such a still. Cooper (1969a, 1973a) has proposed a computer simulation for analysing the performance of such a still. Frick (1970) has also proposed a mathematical model for the still based on the thermic circuits and the Sankey diagrams; his analysis is based on the assumption of sine wave heat flow. Hirschmann and Roefler (1970) have considered periodic insolation in estimating the effect of heat capacity on the performance of the still. The periodic and transient analyses have also been presented by Baum *et al.* (1970), Nayak *et al.* (1980) and Sodha *et al.* (1980), respectively. Apart from the common basin-type solar stills, several other types of stills have also been recently proposed and studied, viz.,

1. multiple-effect stills (Oltra, 1972; Bartali, 1976)
2. tilted tray or inclined-stepped solar stills (Howe, 1961; Akhtamov *et al.*, 1978)
3. tilted, wick type and multiple wick type solar stills (Frick and Somerfeld, 1973; Sodha *et al.*, 1980b; Moustafa, 1979)
4. solar film covered still and wiping spherical stills (Norov *et al.*, 1975; Umarov *et al.*, 1976; Menguy *et al.*, 1980)
5. solar still greenhouse combination (Selcuk, 1970, 1971; Sodha *et al.*, 1980b)
6. indirectly heated solar stills (Soliman, 1976; Malik *et al.*, 1973, 1978; Sodha *et al.*, 1981).

Depending upon their expected life span and application, the various solar stills are categorised into "permanent" (e.g. glass covered stills), "semi-permanent" (e.g. plastic stills) and "expandable" (e.g. double-tube and floating stills) type solar stills; it is expected that the last variety would ultimately prove very inexpensive and easy to handle and transport.

1.3. BASIC PRINCIPLES

A conventional solar still is simply an airtight basin, usually made out of galvanized iron sheet in rectangular shape. It has a top cover of any transparent material, e.g. glass, and the interior surface of its base is blackened to enable absorption of solar energy to the maximum possible extent. Brackish or saline water is poured into the still to fill it partially, which is then exposed to the sun. The glass cover permits solar-radiation to get into the still, which is absorbed predominantly by the blackened base. Consequently, the water gets heated up and hence the moisture content of the air trapped between the water surface and the glass cover increases. The base also radiates energy in the infra-red region which is reflected back into the still by the glass cover; glass is not transparent in the long wavelength region. Thus, the glass cover traps the solar energy inside the still; it also reduces the convective heat losses. The glass cover is usually sloped on one side to enable the water vapor, which condenses on the interior surface, to trickle into a collector.

The most important parameter affecting the output of a solar still is, obviously, the intensity of the solar radiation incident on the still. If Q_t (in Joules/m² day) is the amount of solar energy incident on the glass cover of a still and Q_e (in Joules/m² day) is the energy utilized in vaporizing water in the still, then the daily output of distilled water M_e (in Kg/m² day) is given by

$$M_e = Q_e/\mathcal{L}$$

where $\mathcal{L}$ (in Joules/Kg) is the latent heat of vaporization of water. The efficiency η of the still is given by

$$\eta = \frac{Q_e}{Q_t}$$

It is worthwhile to note that the efficiency of a typical basin-type solar still is quite low and is not greater than 35%.

Figure 1.2 illustrates the principal energy exchange mechanisms in a basin-type solar still. A very large part of the solar radiation, direct and diffuse, falling on the still is absorbed in the blackened base. Small reflection losses occur at the glass surface, the water surface and to a very small extent at the base. The energy absorbed at the base is largely transferred to the water in the still and a small fraction of it is lost to the ground by conduction through the base. Energy is transferred from the water to the glass cover principally by the water vapor evaporating from the water surface and then losing its heat of vaporization to the glass cover upon condensation. Heat is also transferred to the glass cover from water by free convection of the trapped air in the still. The glass cover absorbs a part of the heat radiated from the water surface. A small part of the incident solar energy is also absorbed by the glass cover. The heat thus absorbed by the glass cover is lost to the atmosphere by convection and radiation. Energy exchange also occurs on account of change of sensible heat content of the saline water entering the still, the distillate leaving the still and the brine that accumulates in the still. Thermal losses may also occur due to the leakage of water vapor and the water from the still. A study of the basin-type solar still must take these factors into account. Thus, while the incoming energy is

(1) Solar radiation and
(2) Atmospheric radiation

the outgoing energy comprises of

(1) Convection to atmosphere
(2) Radiation to atmosphere
(3) Reflection to atmosphere
(4) Ground conduction
(5) Edge conduction and convection
(6) Vapor leakage
(7) Brine leakage from basin
(8) Sensible heat of condensate and overflow.

Figure 1.2 illustrates the various components of the energy balance and their direction.

The main components of the energy loss for a typical set of parameters are as follows (Bloemer *et al.*, 1961b):

		Per cent of solar radiation
(1)	Evaporation of distillate (efficiency)	31
(2)	Ground and edge heat losses	2
(3)	Solar radiation reflected from still	11
(4)	Solar radiation absorbed by cover	5
(5)	Radiation from basin water to cover	26
(6)	Internal convection	8
(7)	Re-evaporation of distillate and unaccounted-for losses	17
		100

Introduction

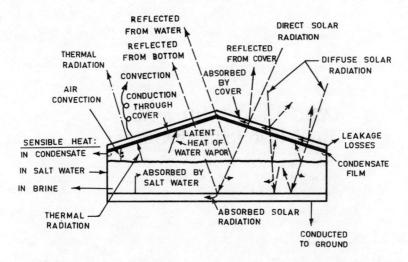

Fig. 1.2. Energy-flow diagram of a basin-type solar still.
(After Lof, 1961b).

From this tabulation, it is evident that the radiation from the basin water to the cover is the largest single heat loss. The heat loss resulting from reflection and absorption of solar radiation by the cover is also important. Re-evaporation is difficult to determine accurately, but calculations and visual inspection indicate that it might constitute a loss of as much as 10% of the available energy. The fact that the ground and edge heat losses are only 2% is especially significant because the bottom of the still was not considered to be insulated.

Study of the deep-basin still, based on the above principles, has shown that the cost of construction can be reduced substantially, while the useful life of the still can at the same time, be increased to possibly 50 years. The improved design requires the use of an asphalt base, concrete support beams, and a glass cover. No partitions need be used between the bays; the supports for the glass cover would rest directly on the asphalt base.

2. BASIC HEAT AND MASS TRANSFER RELATIONS*

The operation of a solar still is governed by various heat transfer modes and therefore a proper understanding of heat transfer is crucial in designing a still. Convection and radiation are the predominant modes of heat transfer in a solar still. A very small amount of energy is also lost to the ground (or atmosphere) due to heat conduction through the base.

It is very convenient to look at the heat transfer modes within the still (hereafter referred to as internal transfer) and between the still and the environment (hereafter referred to as external transfer) separately. It is, however, important to note that the heat transfer problem as such cannot be decoupled. The essential distinction between the modes of heat transfer in these two regions is that, while, within the still convective heat transfer occurs simultaneously with evaporative mass transfer, no such mass transfer occurs outside it. Radiative heat transfer occurs in both the regions along with other modes.

Thermal convection is the process by which heat transfer takes place between a solid surface and the fluid surrounding it. The rate of heat transfer by convection between the fluid and the boundary surface is described by the equation

$$q = h.A.\Delta T$$

which looks disarmingly simple. However, the crux of the problem lies in evaluating the heat transfer coefficient h, which is a complicated function of the geometry of the surface, the flow characteristics of the fluid and the physical properties of the fluid. In most practical cases the heat transfer coefficients are evaluated from empirical equations obtained by correlating experimental results using methods of dimensional analysis. The convective heat transfer is conveniently considered in terms of four dimensionless parameters,** viz. the Nusselt number (Nu), the Grashof number (Gr), the Reynolds number (Re) and the Prandtl number (Pr); the expressions for these

---

*After Cooper (1970).

**For a discussion of significance of these numbers refer to "Heat transfer for engineers", H. Y. Wong, Longman, London, 1977.

numbers are

$$Nu = (h_{cw} x_1 / k_f)$$
$$Gr = (x_1^3 \rho_f^2 g \beta' \Delta T / \mu_f^2)$$
$$Pr = (C_{pf} \mu_f / k_f)$$

and

$$Re = (\rho_f \vartheta_f x_1 / \mu_f)$$

In the above expressions x_1 is the characteristic dimension of the system; in the case of a still x_1 equals the distance between the surface of water and the glass cover.

2.1. INTERNAL TRANSFER

The modes of heat exchange inside the still between the water surface and the glass cover are convection accompanied by evaporative mass transfer (in the form of water vapor) and radiation. In this section, these modes of heat transfer are discussed individually.

2.1.1. *Convection*

Heat is transported across the bulk of the humid air inside the still by free convection[*] of air. It then releases its enthalpy upon coming into contact with the glass cover, which is at a lower temperature. The coefficient of heat transfer is usually incorporated in the Nusselt number. In the case of heat transfer by free convection, the Nusselt number is related to the Grashof and Prandtl numbers i.e.

$$Nu = f(Gr, Pr) \qquad (2.1.1)$$

For heat flow from the horizontal water surface in the upward direction, i.e. against forces of gravity, Jakob (1949, 1957) has arrived at the following relationship by correlating the experimental data of Mull and Reiher

$$Nu = C(Gr.Pr)^n \qquad (2.1.2)$$

The values of C and n in the various range of values of Gr are:

  (i)   for $Gr < 10^3$, $C = 1$, $n = 0$,
 (ii)   for $10^4 < Gr < 3.2 \times 10^5$, $C = 0.21$, $n = 1/4$,
(iii)   for $3.2 \times 10^5 < Gr < 10^7$, $C = 0.075$, $n = 1/3$.

In the first case the magnitude of convection is negligible, in the second, the air flow is laminar and in the last case the air flow is in the turbulent regime.

---

[*]In free convection the fluid motion is caused by the action of buoyancy forces arising from the density variations in the fluid which in turn exists whenever there is a temperature gradient in the fluid. In contrast fluid motion in forced convection arises due to externally applied pressure gradient.

For convective heat transfer by humid air with simultaneous mass transfer of a fluid of lower molecular weight (i.e. water vapor), it becomes necessary to use a special Grashof number Gr'. It has been shown by Sharpley and Boelter (1938), that for non-isothermal evaporation,

$$Gr' = \frac{x_1^3 \rho_f^2 g}{\mu_f^2} \left[\frac{M_\infty T_o'}{M_o T_\infty'} - 1 \right] \tag{2.1.3}$$

where the subscripts o and ∞ refer to the conditions at the surface at which the evaporation originates and to a point far away from this surface. The subscript f refers to air saturated with water vapor. For $M_\infty = M_o$ the term in brackets reduces to $\beta' \Delta T$; $\beta' \approx T_\infty'^{-1}$.

Let subscripts ω and a refer to the diffusing (water vapor) and inert (air) gases respectively. Assuming these to be ideal gases (a necessary condition for the modified equation 2.1.3) and remembering that $P_{\omega\infty} + P_{a\infty} = P_{\omega o} + P_{ao} = P_T$ (total gas pressure), one obtains

$$\frac{M_\infty T_o'}{M_o T_\infty'} - 1 = \frac{M_\omega P_{\omega\infty} T_o' + M_a P_{a\infty} T_o'}{M_\omega P_{\omega o} T_\infty' + M_a P_{ao} T_\infty'} - 1$$

$$= \frac{M_\omega (P_{\omega\infty} T_o' - P_{\omega o} T_\infty') + M_a (P_{a\infty} T_o' - P_{ao} T_\infty')}{T_\infty' (M_\omega P_{\omega o} + M_a P_{ao})}$$

$$= \frac{(M_\omega P_{\omega o} + M_a (P_T - P_{\omega o}))(T_o' - T_\infty') + (M_\omega - M_a)(P_{a\infty} - P_{\omega o}) T_o'}{T_\infty' (M_\omega P_{\omega o} + M_a (P_T - P_{\omega o}))}$$

$$= \frac{(T_o' - T_\infty')}{T_\infty'} + \frac{(M_\omega - M_a)(P_{\omega\infty} - P_{\omega o}) T_o'}{T_\infty' (M_a P_T + (M_\omega - M_a) P_{\omega o})} \tag{2.1.4}$$

For an air-water vapor system at normal atmospheric pressure:

$M_\omega = 18$

$M_a = 28.96$

$P_T = 98.07 \times 10^3$ pa.

Substitution in equation (2.1.4) gives:

$$\frac{M_\infty T_o'}{M_o T_\infty'} - 1 = \frac{(T_o' - T_\infty')}{T_\infty'} + \frac{(P_{\omega o} - P_{\omega\infty}) T_o'}{(268.9 \times 10^3 - P_{\omega o}) T_\infty'}$$

$$= \beta' \left[\Delta T + \frac{(P_{\omega o} - P_{\omega\infty}) T_o'}{(268.9 \times 10^3 - P_{\omega o})} \right]$$

$$= \beta' \Delta T' \tag{2.1.5}$$

where $\Delta T'$ is the equivalent temperature difference. Thus:

$$Gr' = \frac{x_1^3 \rho_f^2 g \beta'}{\mu_f^2} \Delta T' \tag{2.1.6}$$

Basic Heat and Mass Transfer Relations

which is the same relationship as that used by Dunkle (1961).

For a mean air temperature of 50°C and an equivalent temperature difference of 17°C and assuming saturated air, the special Grashof number given by equation (2.1.6) is $2.81 \times 10^7 \cdot x_1^3$. Over the normal operating temperature range and for values of the mean water to cover spacing (x_1) not too small, the special Grashof number may be seen to lie in the range for which $C = 0.075$ and $n = \frac{1}{3}$ in equation (2.1.2). For this value of the exponent n, the length parameter conveniently cancels out so that the heat transfer coefficient is substantially independent of the spacing x_1. Dunkle (1961) has chosen those values of the physical parameters (occurring in the dimensionless variables) which are applicable over the normal range of operation of the still, to arrive at the relationship:

$$q_{c\omega} = 0.884 \left[T_\omega - T_{gi} + \frac{(P_\omega - P_g)(T_\omega + 273)}{(268.9 \times 10^3 - P_\omega)} \right] \cdot (T_\omega - T_{gi}) \qquad (2.1.7(a))$$

$$= h_{c\omega}(T_\omega - T_{gi}) \qquad (2.1.7(b))$$

2.1.2. Evaporation

The mass of air transferred per unit area per unit time by free convection is given by:

$$\dot{M}_a = \frac{q_{c\omega}}{C_{pa}(T_\omega - T_{gi})} = \frac{h_{c\omega}}{C_{pa}} \qquad (2.1.8)$$

It is assumed that the air next to the water surface is saturated at the water temperature and therefore, the specific humidity (or mass of water per unit mass of dry air) may be written as

$$\frac{M_\omega}{M_a} \cdot \frac{P_\omega}{(P_T - P_\omega)} \qquad (2.1.9)$$

Thus, the mass of water vapor transferred per unit area per unit time from the water surface is:

$$\frac{M_\omega}{M_a} \cdot \frac{P_\omega}{(P_T - P_\omega)} \cdot \frac{h_{c\omega}}{C_{pa}} \qquad (2.1.10)$$

Similarly, the mass of water vapor transferred per unit area per unit time from the glass surface is given by

$$\frac{M_\omega}{M_a} \cdot \frac{P_g}{(P_T - P_g)} \cdot \frac{h_{c\omega}}{C_{pa}} \qquad (2.1.11)$$

Baum (1964) found that the bulk of the air did not have any significant rate of exchange of water vapor with the boundary layers at the water and glass surface. Then the net mass of water vapor transferred per unit area per unit time is given by the difference of the expressions (2.1.10) and (2.1.11), i.e.

$$\frac{M_\omega}{M_a} \cdot \frac{h_{c\omega}}{C_{pa}} \left[\frac{P_\omega}{(P_T-P_\omega)} - \frac{P_g}{(P_T-P_g)} \right] \qquad (2.1.12)$$

The rate at which heat is transferred from the water surface to the glass cover on account of the mass transfer of the water vapor is:

$$q_{e\omega} = \frac{M_\omega}{M_a} \cdot \frac{(P_\omega-P_g)}{C_{pa}} \cdot \mathcal{L} \cdot h_{c\omega} \cdot \frac{P_T}{(P_T-P_\omega)(P_T-P_g)} \qquad (2.1.13(a))$$

$$= h_e(P_\omega-P_g) \qquad (2.1.13(b))$$

In a practical situation P_ω and P_g are considerably smaller than P_T and therefore, one is justified in assuming that $(P_T-P_\omega)(P_T-P_g)$ is approximately equal to P_T^2. The equivalent mass transfer coefficient h_e may then be written in terms of the heat transfer coefficient $h_{c\omega}$ as

$$\frac{h_e}{h_{c\omega}} = \frac{\mathcal{L}}{C_{pa}} \cdot \frac{M_\omega}{M_a} \cdot \frac{1}{P_T} \qquad (2.1.14)$$

The above equation can also be obtained by using the Lewis' relation (1922, 1933), i.e.

$$\frac{h_{c\omega}}{h_D \rho_a C_{pa}} = 1 \; ; \qquad (2.1.15)$$

in other words the ratio of the heat transfer coefficient to the mass transfer coefficient is equal to the specific heat per unit volume at constant pressure of the mixture; h_D is defined such that the mass transfer rate per unit area from the water surface is

$$\frac{\dot{m}}{A} = h_D(\rho_\omega-\rho_a) \qquad (2.1.16)$$

where $\dot{m}$ = mass transfer rate, Kg/hr.
h_D = mass transfer coefficient, Kg per (hour) (square meter)/(Kg per meter^3)
ρ = partial mass density of water vapor, Kg/m^3

By using the perfect gas equation $\{PV = RT', (V = \frac{M}{\rho})\}$ for water vapor and substituting h_D from equation (2.1.15) in equation (2.1.16), one obtains

$$\frac{\dot{m}}{A} = \frac{h_{c\omega}}{\rho_a C_{pa}} \cdot \frac{M_\omega}{RT'} (P_\omega-P_a) \; ;$$

we have assumed that $T' \approx T'_\omega \approx T'_a$, which is a reasonable approximation.

Now the amount of heat transferred on account of the mass transfer of the water vapor per unit area per unit time is, hence

$$q_{e\omega} = \frac{\dot{m}\mathcal{L}}{A} = \mathcal{L} \cdot \frac{h_{c\omega}}{\rho_a C_{pa}} \cdot \frac{M_\omega}{RT'} (P_\omega-P_a)$$

$$= h_e(P_\omega-P_a)$$

where

$$\frac{h_e}{h_{c\omega}} = \frac{\mathcal{L}}{\rho_a C_{pa}} \cdot \frac{M_\omega}{RT'}$$

Using the perfect gas equation for air, i.e.

$$RT' = \frac{M_a P_a}{\rho_a} \quad (P_a = \text{atmospheric pressure})$$

$$\frac{h_e}{h_{c\omega}} = \frac{\mathcal{L}}{C_{pa}} \cdot \frac{M_\omega}{M_a} \cdot \frac{1}{P_T} \tag{2.1.17}$$

because $P_T \approx P_a$ for small P_ω as discussed earlier.

The above equation is identical to Eq. (2.1.14). Substitution of the appropriate values for the different parameters yields:

$$h_e = 0.013\, h_{c\omega} \tag{2.1.18}$$

Note that in the formulation of equations (2.1.8) to (2.1.14), the water vapor is assumed to obey the perfect gas equation and consequently the gas constant R is a universal constant.

The value of $h_e/h_{c\omega}$ as given by equation (2.1.18) viz. 0.013, is smaller than that obtained by Bowen (1926) and Dunkle (1961). This is due to the approximation made earlier in expanding the factor $(P_T - P_\omega)(P_T - P_g)$ and neglecting terms involving P_ω and P_g. It is seen that the best representation of the mass-heat transfer phenomena is obtained if the value of $h_e/h_{c\omega}$ is taken to be 16.273×10^{-3}. Thus the heat transferred per unit area per unit time by evaporation from the water surface to the glass cover is

$$q_{e\omega} = 16.273 \times 10^{-3}\, h_{c\omega}(P_\omega - P_g) \tag{2.1.19}$$

and the mass transfer rate $\dot{m}_e$ is:

$$\dot{m}_e = \frac{q_{e\omega}}{\mathcal{L}} \tag{2.1.19(a)}$$

Convective heat transfer from basin liner to water mass has been discussed in Appendix B.

2.1.3. Radiation

In the usual analyses of solar stills, the water surface and the glass cover are considered as infinite parallel planes; this is a valid approximation for stills with small cover slopes and large dimensions in both the horizontal directions. This would not be valid for stills consisting of essentially unconnected strips or long rectangles.

For the infinite case:

$$q_{r\omega} = \frac{\sigma(T_1^4 - T_2^4)}{\frac{1}{\epsilon_1} + \frac{1}{\epsilon_2} - 1} \tag{2.1.20}$$

Absolute values of the total energy transfer rate are obtained by the addition of equations (2.1.7), (2.1.19) and (2.1.20). To provide an insight into the relative magnitudes of each mode, the heat transferred by evaporation, radiation and convection, expressed as a fraction of the total, is shown in Fig. 2.1 for various water temperatures and water-glass temperature differences. The important characteristic of Fig. 2.1 is that, for high mean temperatures of operation, the evaporation fraction is large and hence the internal efficiency is high.

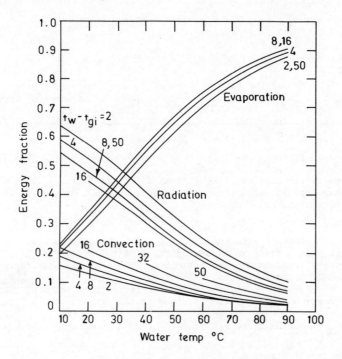

Fig. 2.1. Evaporative, radiative and convective energy fractions. (After Cooper, 1973b).

2.2. EXTERNAL TRANSFER MODE

2.2.1. *Convection and Radiation*

Due to the small thickness of the glass cover, the temperature in the glass may be assumed to be uniform. The external radiation and convection losses from the glass cover to outside atmosphere can be expressed as,

$$q_{ra} = \epsilon_g \sigma \cdot [(T_g+273)^4 - (T_{sky}+273)^4] , \qquad (2.2.1)$$

and

$$q_{ca} = h_{ca}(T_g-T_a) \qquad (2.2.2)$$

and hence,

$$q_a = q_{ra} + q_{ca} ; \qquad (2.2.3)$$

$T_{sky} = (T_a - 12)$ is the apparent sky temperature for long wave radiation exchange, assumed to be 12°C below ambient; Eg for glass ≈ 0.85. The external convection coefficient h_{ca} is a function of wind velocity and is given below (Duffie, 1974),

$$h_{ca} = 5.7 + 3.8 \, \vartheta \qquad (2.2.4)$$

2.2.2. Periodic Heat Transfer in Conducting Media

Energy transfer through the base of the still occurs through thermal conduction. Since the solar insolation is a periodic phenomenon, the heat transfer into the ground is also of periodic nature. We now proceed to briefly outline a mathematical approach to study the periodic heat transfer through a conducting medium.

The temperature distribution $\theta(x,t)$ in a uniform medium is described by the well known Fourier's equation of heat conduction viz.

$$\frac{K}{\rho_o C} \cdot \nabla^2 \theta(x,t) = \frac{\partial \theta(x,t)}{\partial t} \qquad (2.2.5)$$

In the analysis of desalination we are mainly concerned with the situations in which the temperature varies only along one direction (to be specific the x axis); thus

$$\frac{K}{\rho_o C} \frac{\partial^2 \theta}{\partial x^2} = \frac{\partial \theta}{\partial t} \qquad (2.2.6)$$

If the boundary conditions (to be discussed later) are periodic, the temperature distribution is also periodic in the steady state and may in general be written as

$$\theta(x,t) = \theta_o(x) + \mathrm{Re} \sum_{n=1}^{\infty} \theta_n(x) \exp(in\omega t) \qquad (2.2.7)$$

where R_e denotes the real part.

Substituting for $\theta(x,t)$ from Eq. (2.2.7) in Eq. (2.2.6), one obtains

$$\frac{d^2 \theta_o}{dx^2} = 0 \qquad (2.2.8(a))$$

and

$$\frac{d^2 \theta_n}{dx^2} = \beta_n^2 \theta_n \qquad (2.2.8(b))$$

where

$$\beta_n = \left(\frac{in\omega\rho_o c}{K}\right)^{\frac{1}{2}} = n^{\frac{1}{2}}\left(\frac{\omega\rho_o c}{2K}\right)^{\frac{1}{2}}(1+i)$$

$$= \alpha\sqrt{n}\,(1+i)$$

$$= \alpha_n + i\,\alpha_n \qquad (2.2.8(c))$$

The general solution of the above equations is

$$\theta_o = Ax + B, \qquad (2.2.9(a))$$

and

$$\theta_n = C_n \exp(-\beta_n x) + D_n \exp(B_n x) \qquad (2.2.9(b))$$

From Eqs. (2.2.7) and (2.2.9), the steady state periodic distribution of a temperature in a uniform conducting medium may be written as

$$\theta(x,t) = Ax + B + \mathrm{Re}\sum_{n=1}^{\infty}[C_n \exp(-\beta_n x) + D_n \exp(\beta_n x)]\cdot \exp(in\omega t)$$

$$= (Ax + B) + \sum_{n=1}^{\infty} C_{on}\exp(-\alpha_n x)\cdot \cos(n\omega t - \alpha_n x + \varepsilon_n)$$

$$+ \sum_{n=1}^{\infty} D_{on}\exp(\alpha_n x)\cdot \cos(n\omega t + \alpha_n x + \varepsilon_n')$$

where*

$$C_n = C_{on}\exp(i\varepsilon_n) \text{ and } D_n = D_{on}\exp(i\varepsilon_n')$$

2.2.3. *Periodic Boundary Conditions*

1. Interface of two conducting media: The temperature θ and the heat flux $(-K\frac{\partial \theta}{\partial x})$ across the interface (say x=L) are continuous.

2. Semi-infinite media: The temperature θ is finite as $x \to \infty$, hence $A = o$ and $D_n = o$.

---

*Some care has to be exercised in expressing a complex number $A = a + ib$ in the form $A_o \exp(i\varepsilon)$
If $\tan^{-1}(b/a) = \varepsilon'(|\varepsilon'| < \pi/2)$
  $\varepsilon = \varepsilon'$ when a is positive
and $\varepsilon = \pi + \varepsilon'$ when a is negative;
A_o is of course equal to $\sqrt{(a^2 + b^2)}$

3. The temperature on one of the boundaries (say x = L) is periodic i.e.

$$\theta(L,t) = \theta'_o + \text{Re} \sum_{n=1}^{\infty} \theta'_n \exp(in\omega t)$$

4. The heat flux on and heat losses from a boundary may be periodic.

3. SINGLE BASIN SOLAR STILL

3.1. BACKGROUND

A single basin solar still is characterized by a single basin having saline water; it may, of course, have more than one transparent cover. Schematic diagrams of some of the common designs of single basin still are shown in Figs. 3.1. These designs differ in structure and materials of construction, but basically, incorporate common elements for different functions. The glass/plastic cover is transparent to the incoming solar radiation, but blocks the long wavelength radiation emitted by the water surface and the base of the still.* Besides this, the cover prevents the escape of the humid air trapped inside the still and also provides a cool surface for condensation of water vapor. The cover should be sloped on one side at an angle large enough to facilitate an easy flow of the water droplets, that condense on it, into a condensate trough. The trough is also inclined so that the collected water flows out of the still to be collected in bottles. Obviously, the condensate trough should extend all along the lower edge of the sloped cover. It is important to note that the angle at which the cover is inclined should not be so large as to present a grazing surface to the solar rays at noon, when the solar intensity is maximum; this would cut off a very large fraction of the total daily insolation.

The basin of the still should be watertight and the entire still should be airtight, except for the inlets and outlets. The base of the still should be blackened on its interior surface to enable maximum absorption of solar radiation. Cooper (1973b) suggested that the interior surface of the walls of the still may be painted glossy white with a brushable grade of silicone in order to reflect solar radiation incident on them onto the saline water and also to protect the surface from corrosion. However, the white surface is cool enough for condensation of water vapor; this condensate trickles down into the basin, which is not desirable. The still may be placed on the ground with a thin layer of insulation separating it from the ground

---

*Neither glass nor plastic sheets are transparent in the entire wavelength range of the solar radiation. However, the amount of energy reflected or absorbed by them is in practice seen to be insignificant.

(Fig. 3.2a); this still is called the ground still. On the other hand it may be mounted in a box (usually made of wood) such that it is well insulated on all sides except the top to reduce heat losses to a minimum (Fig. 3.2b). Such a still is called a mounted still.

Use of durable materials such as metals, concrete etc. in constructing the stills increases their life. Table 3.1 contains some interesting data on several major solar distillation plants. Details of a typically large distillation plant viz. the one at Awania, Gujarat in India are given in Appendix A.

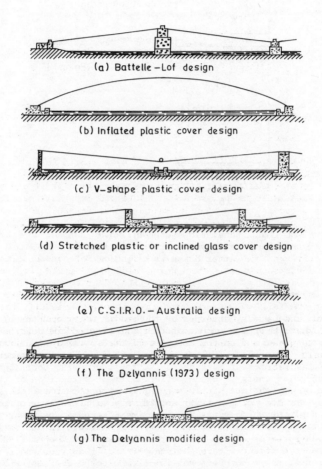

Fig. 3.1. (After, Delyannis and Delyannis, 1973.)

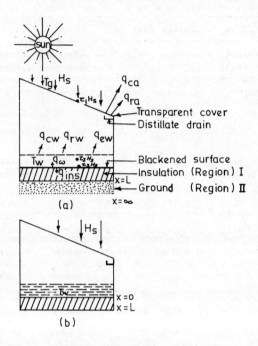

Fig. 3.2. Schematic sketch of solar still.
(a) Ground still; (b) mounted still.
(After Nayak et al., 1980.)

3.2. SOLAR RADIATION BALANCE

The amount of solar energy absorbed by the saline water and the blackened base of the still depends on the amount of energy reflected and absorbed by the transparent top cover and on the amount reflected by the saline water surface and the base of the still. The water film formed due to the condensation of water vapor on the interior surface of the top cover also contributes to the reflection of solar energy. Since reflectivity is a function of the angle of incidence on any surface, it is natural to expect that the energy absorption by the base of the still should vary with the time of the day, because the angle at which sun's rays are incident on a stationary still varies with time. Fortunately such temporal variations in the reflectivities at the various surfaces of the still are quite small. Cooper (1970) has shown that, for places confined within latitudes 0 and 45 degrees and for the angle of inclination of the top covers ranging from 0 to 60 degrees, the variation with time in the energy absorbed by the saline water and the base of the still is very small. Consequently, the reflection and transmission coefficients may be treated as constants for any particular still. In what follows the average values (with respect to time) of these coefficients at the various surfaces of the still will be assumed to be constant in time for a particular still. A glance at Table 3.2 would help one appreciate the approximation involved in making this assumption.

Single Basin Solar Still

TABLE 3.1. Some Solar Distillation Plants*

Country	Location	Design	Year	m²	Feed	Cover	Remarks
Australia	Muresk I		1963	372	Brackish	Glass	Rebuilt
	Muresk II		1966	372	Brackish	Glass	Operating
	Coober Pedy	3.1e	1966	3160	Brackish	Glass	Operating
	Caiguna		1966	372	Brackish	Glass	Operating
	Hamelin Pool		1966	557	Brackish	Glass	Operating
	Griffith		1967	413	Brackish	Glass	Operating
Cape Verde Isl	Santa Maria	3.1c	1965	743	Seawater	Plastic	
	Santa Maria		1968				Abandoned
Chile	Las Salinas	3.1e	1872	4460	Brackish	Glass	Abandoned
	Quillagua		1968	100	Seawater	Glass	Operating
Greece	Symi I	3.1d	1964	2686	Seawater	Plastic	Rebuilt
	Symi II	3.1d	1968	2600	Seawater	Str.Plas.	Dismantled
	Aegina I	3.1c	1965	1490	Seawater	Plastic	Rebuilt
	Aegina II	3.1d	1968	1486	Seawater	Str.Plas.	Abandoned
	Salamis	3.1c	1965	388	Seawater	Plastic	Abandoned
	Patmos		1967	8600	Seawater	Glass	Operating
	Kimolos		1968	2508	Seawater	Glass	Operating
	Nisyros		1969	2005	Seawater	Glass	Operating
	Fiskardo	3.1f	1971	2200	Seawater	Glass	Operating
	Kionion		1971	2400	Seawater	Glass	Operating
	Megisti		1973	2528	Seawater	Glass	Operating
India	Bhavnagar	3.1e	1965	377	Seawater	Glass	Operating
	Awania	3.1e	1978	1866	Brackish	Glass	Operating
Mexico	Natividad Isl	3.1d	1969	95	Seawater	Glass	Operating
Pakistan	Gwadar I	3.1f	1969	306	Seawater	Glass	Operating
	Gwadar II	3.1g	1972	9072	Seawater	Glass	Operating
Spain	Las Marinas	3.1a	1966	868	Seawater	Glass	Operating
Tunisia	Chakmou	3.1d	1967	440	Brackish	Glass	Operating
	Mahdia		1968	1300	Brackish	Glass	Operating
U.S.A.	Daytona Beach	3.1a	1959	228	Seawater	Glass	Rebuilt
	Daytona Beach		1961	246	Seawater	Glass	Dismantled
	Daytona Beach	3.1b	1961	216	Seawater	Plastic	Dismantled
	Daytona Beach		1963	148	Seawater	Plastic	Dismantled
USSR	Bakharden	3.1e	1969	600	Brackish	Glass	Operating
West Indies	Potit St.Vincent	3.1b	1967	1710	Seawater	Plastic	Operating
	Haiti	3.1d	1969	223	Seawater	Glass	Operating
India	Bitra	3.1c	1980	–	Brackish	Glass	Operating (capacity 2000 l/day)
	Kulmis		1980	–	Brackish	Glass	Operating (capacity 3000 l/day)
China	Wuzhi	3.1c	1976	385	Seawater	Glass	Operating
	Zhungjian		1979	50	Seawater	Glass	Operating

*The designs correspond to Fig. 3.1. After Delyannis and Delyannis, (1973).

TABLE 3.2. Effect of Angle of Incidence on Radiation Parameters; the values are expressed as percent fractions*

Angle of incidence, deg.	0	30	45	60
Cover reflection	5	5	6	10
Cover absorption	5	5	5	5
Cover transmission	90	90	89	85
Water reflection	2	2	3	6
Water absorption	30	30	30	30
Water transmission	68	68	67	64
Basin reflection	5	5	5	5
Basin absorption	95	95	95	95
Basin transmission	0	0	0	0

*After Cooper (1969b).

3.3. ANALYSIS OF AN IDEAL STILL

An ideal solar still is one with zero conductive heat losses and a saline water level shallow enough to enable one to neglect the sensible heat stored in the saline water as compared to the energy transferred to and from the saline water. Consequently, an ideal still would attain a steady state almost instantaneously corresponding to the prevailing circumstances. Cooper (1973b) has studied the performance of such a still.

The interior heat transfer rates (i.e. heat transfer rates inside the still) $q_{c\omega}$, $q_{e\omega}$ and $q_{r\omega}$ are given by Eqs. (2.1.7), (2.1.19) and (2.1.20), respectively. The solar radiation absorbed by the top cover, together with the heat transferred to it by the condensing water vapor, is rejected to the atmosphere through convection and radiation; the convection would be a strong function of the wind velocity outside the still. Under steady state conditions, the heat balance on the saline water surface may be expressed as

$$\alpha_\omega H_s = (q_{c\omega} + q_{e\omega} + q_{r\omega}) + \frac{C_{p\omega}}{\ell}(T_\omega - T_a)q_{e\omega} \qquad (3.3.1)$$

In writing the above equation it is assumed that, a steady inflow of saline water at temperature T_a replaces the water lost through evaporation. Thus, the second term on the right hand side of this equation represents the heat used up in raising the temperature of the incoming saline water from T_a to T_ω, the steady state temperature of the saline water in the still.

In order to relate the temperature on the interior surface of the top cover, T_{gi}, to the ambient temperature T_a, it is assumed, as a first approximation, that the temperature profile through the top cover is linear and is independent of the solar radiation absorbed by the cover. If the thickness of the cover is small, this absorption may be considered as having occurred at the external surface of the cover. One may then write

$$(q_{c\omega} + q_{e\omega} + q_{r\omega}) = \frac{K_g}{\ell_g}(T_{gi} - T_{go}) \qquad (3.3.2)$$

where T_{gi} and T_{go} are the temperatures of the interior and exterior surfaces of the transparent cover. The external radiative and convective heat transfer rates, q_{ca} and q_{ra} are given by Eqs. 2.2.1 and 2.2.2. Clubbing all these equations together, the total energy balance equation may be written as

Single Basin Solar Still

$$0.85 \sigma [(T_{go}+273)^4 - (T_a+261)^4] + h_{ca}(T_{go}-T_a)$$

$$= \alpha_g H_s + \frac{K_g}{\ell_g}(T_{gi}-T_{go}) \qquad (3.3.3)$$

Although no explicit solution exists for this equation, one may always calculate the distillation rate approximately by iteration. Based on such a simplified model, Cooper (1973b) has shown that the efficiency of an ideal solar still may be as high as 60% for high values of solar insolation.

3.4. PERIODIC THEORY OF SOLAR STILL

The steady state theory of solar stills is, at best, an approximation of doubtful validity because solar insolation and ambient air temperature are periodic parameters with a period of 24 hours. It is, therefore, very necessary to incorporate periodicity in the analysis of a solar still. This fact is borne out by the investigations (among others) of Cooper (1969a, 1973a), Frick (1970), Hirschmann and Roefler (1970) and Baum *et al.* (1970). A major lacunae in all these investigations is that, in the Fourier expansion of all the periodic quantities, only the fundamental harmonics are retained. This is not an accurate representation of the solar insolation and ambient temperature; one has to include a few higher order terms.

The other major approximation made in the analysis of an ideal solar still can also be profitably dropped. A small amount of heat conduction always occurs despite good thermal insulation; thus, explicit solution of the heat conduction equation also becomes necessary. In this section, following Nayak *et al.* (1980), the performance of a solar still with periodic solar insolation and ambient temperature is presented taking into account the appropriate heat and mass transfer relations (section 2) and the heat lost by conduction into the ground and the atmosphere.

3.4.1(a). Ground Still

A solar still placed directly on the ground is called a ground still. A schematic of the basic configuration of such a still is shown in Fig. 3.2(a). The figure also shows the basic heat flux components at the various surfaces. The energy balance conditions at the top cover, the saline water surface and the absorbing surface may be written as

$$M_g \frac{dT_g}{dt} = \tau_1 H_s + (q_{r\omega}+q_{c\omega}+q_{e\omega}) - q_a \qquad (3.4.1)$$

$$M_{\omega o} \frac{dT_\omega}{dt} = \tau_2 H_s + q_\omega - (q_{r\omega}+q_{c\omega}+q_{e\omega}) \qquad (3.4.2)$$

$$\tau_3 H_s = q_\omega + q_{ins} \qquad (3.4.3)$$

respectively. The heat transfer rates q_a, $q_{c\omega}$, $q_{e\omega}$ and $q_{r\omega}$ are defined by Eqs. (2.2.3), (2.1.7), (2.1.19) and (2.1.20) respectively, and are the same as those defined in section 2. In writing the above heat balance conditions, the following assumptions have been made.

1. The amount of water lost through evaporation is small compared to the amount of saline water in the basin. This is reasonable if the depth of the water column in the basin is large. If the water column is shallow then the water level may be maintained constant by a steady inflow of saline water at such a rate as to replace the amount of water lost through evaporation. Thus, the mass of saline water in the basin is assumed to be constant in any situation. Further, it is assumed that the heat required to raise the temperature of a particular quantity of the inflowing saline water from ambient to that of the water inside the still is negligible compared to the heat required to evaporate the same quantity of water, i.e.

$$C_{p\omega}(T_\omega - T_a) << \mathcal{L}$$

This is valid to a good approximation.

2. There is no leakage of water vapor from the still.

3. No temperature gradients exist along the vertical direction in either the top cover or the water in the basin.

4. The surface areas of the top cover, the water surface and the base of the still are equal.

Under normal operating conditions the rise in temperature of the top cover and the saline water in the still is small and within this temperature range the vapor pressure inside the still may be approximately written as a linear function of temperature (Sodha et al., 1978), i.e.

$$P = R_1 T + R_2 \quad (3.4.4)$$

where the constants R_1 and R_2 may be evaluated by fitting the saturation vapor pressure data in the temperature range of interest to a straight line. Thus, the heat transfer rates inside the still become

$$q_{c\omega} = h_{c\omega}(T_\omega - T_g) \quad (3.4.5)$$

$$q_{e\omega} = h_{eff}(T_\omega - T_g) \quad (3.4.6)$$

and

$$q_{r\omega} = h_{r\omega}(T_\omega - T_g) \quad (3.4.6(a))$$

where the heat transfer coefficients $h_{c\omega}$, h_{eff} and $h_{r\omega}$ are written as

$$h_{c\omega} = 0.884 \left[T_\omega - T_g + \frac{R_1(T_\omega - T_g)(T_\omega + 273)}{268.9 \times 10^3 - R_2 - R_1(T_\omega + 273)} \right]^{1/3}, \quad (3.4.7(a))$$

$$h_{eff} = 16.276 \times 10^{-3} h_{c\omega} R_1 \quad (3.4.7(b))$$

and

$$h_{r\omega} = \frac{\epsilon \sigma [(T_\omega + 273)^4 - (T_g + 273)^4]}{(T_\omega - T_g)} \quad (3.4.7(c))$$

Single Basin Solar Still

It should be mentioned that the variation of the latent heat of vaporization with temperature has been neglected. It is evident from Eqs. (3.4.7), that the heat transfer coefficients are temperature dependent quantities, but they may be treated as constants within the small range over which the temperature changes occur in the top cover and the saline water. Thus, one may calculate these coefficients from Eqs. (3.4.7) by replacing T_ω and T_g by their mean values.

The heat transfer rates outside the still may similarly be written as

$$q_a = h_2(T_g - T_a) \quad , \tag{3.4.8}$$

where

$$h_2 = h_{ca} + \epsilon_g \sigma \frac{[(T_g + 273)^4 - (T_a + 261)^4]}{T_g - T_a} \tag{3.4.9}$$

and

$$h_{ca} = 5.7 + 3.8\vartheta \quad \text{(Duffie, 1974)} \tag{3.4.9(a)}$$

In Eq. (3.4.9) too, mean values of T_g and T_a may be substituted to calculate approximate h_2, which is to be treated as a constant.

Energy transferred from the absorbing surface to the saline water can be written as

$$q_\omega = h_3(\theta_I|_{x=0} - T_\omega) \tag{3.4.10}$$

while the loss to the insulation from the absorbing surface is given by

$$q_{ins} = -K_1\left(\frac{\partial \theta_I}{\partial x}\right)_{x=0} \tag{3.4.11}$$

where $\theta_I(x,t)$ is the temperature distribution below the absorbing surface.

Equations (3.4.1)-(3.4.3) can now be rewritten as

$$M_g \frac{dT_g}{dt} = \tau_1 H_s + h_1(T_\omega - T_g) - h_2(T_g - T_a) \quad , \tag{3.4.12}$$

$$M_\omega \frac{dT_\omega}{dt} = \tau_2 H_s + h_3(\theta_I|_{x=0} - T_\omega) - h_1(T_\omega - T_g) \quad , \tag{3.4.13}$$

and

$$\tau_3 H_s = h_3(\theta_I|_{x=0} - T_\omega) - K_1\left(\frac{\partial \theta_I}{\partial x}\right)_{x=0} \quad , \tag{3.4.14}$$

where,

$$h_1 = h_{r\omega} + h_{c\omega} + h_{eff}$$

The temperature distribution $\theta(x,t)$ in the insulation and ground is governed by the heat conduction equation

$$\rho_j C_j \frac{\partial \theta}{\partial t} = K_j \frac{\partial^2 \theta}{\partial x^2} \quad , \qquad j = 1,2 \qquad (3.4.15)$$

subject to the boundary conditions (Fig. 3.2a)

$$\theta_I(x=L,t) = \theta_{II}(x=L,t) \quad , \qquad (3.4.16(a))$$

$$-K_1 \left(\frac{\partial \theta_I}{\partial x}\right)_{x=L} = - K_2 \left(\frac{\partial \theta_{II}}{\partial x}\right)_{x=L} \quad , \qquad (3.4.16(b))$$

and

$$\theta_{II}(x,t) \text{ is finite for } x \to \infty \qquad (3.4.16(c))$$

The solar intensity and ambient air temperature can be considered to be periodic in time and hence can be Fourier analysed in the form

$$H_s = a_o + \text{Re} \sum_n a_n \exp(in\omega t) \qquad (3.4.17(a))$$

and

$$T_a = b_o + \text{Re} \sum_n b_n \exp(in\omega t) \qquad (3.4.17(b))$$

where,

$$a_n = A_n \exp(-i\sigma_n) \quad ,$$

$$b_n = B_n \exp(-i\phi_n) \quad ,$$

$$\omega = 2\pi/(24 \times 60 \times 60) \cdot \sec^{-1}$$

a_o, b_o, A_n, B_n, σ_n and ϕ_n are given in Tables 3.3 and 3.4 respectively for a typical hot day in Delhi.

TABLE 3.3. Fourier Coefficients of Solar Intensity Available on the Glass Surface (16 March, 1979 New Delhi)

n	0	1	2	3	4	5	6
A_n (W/m²)	213.056	342.54	166.24	47.969	39.583	16.097	6.614
σ_n (in degrees)	-	179.01	351.28	120.842	239.19	63.23	320.12

TABLE 3.4. Fourier Coefficients of Ambient Air Temperature
(16 March, 1979 New Delhi)

n	0	1	2	3	4	5	6
B_n (°C)	23.8833	5.2754	1.4713	0.3163	0.1741	0.13	0.03322
ϕ_n (in degrees)	-	216.906	317.233	15.071	30.15	28.119	124.012

In view of Eqn. (3.4.7) one can assume the following periodic solutions (McAdams, 1954; Threlkeld, 1970):

$$\theta_I(x,t) = [A_1 x + B_1 + \text{Re} \sum_n \{C_{1n} \exp(-\beta_{1n} x) + D_{1n} \exp(\beta_{1n} x)\} \times \exp(in\omega t)], \text{ for } 0 \le x \le L \quad (3.4.18(a))$$

$$\theta_{II}(x,t) = B_2 + \text{Re} \sum_n C_{2n} \exp(-\beta_{2n} x) \exp(in\omega t), \text{ for } x \ge L \quad (3.4.18(b))$$

$$T_g(t) = g_o + \text{Re} \sum_n g_n \exp(in\omega t) \quad (3.4.18(c))$$

and

$$T_\omega(t) = H_o + \text{Re} \sum_n H_n \exp(in\omega t) ; \quad (3.4.18(d))$$

where,

$$\beta_{1n} = \sqrt{n}\, \alpha_1 (1+i) ,$$

$$\beta_{2n} = \sqrt{n}\, \alpha_2 (1+i) ,$$

$$\alpha_j = \sqrt{\frac{\omega P_j C_j}{2 K_j}} , \quad j = 1,2$$

It may be noted that Eqn. (3.4.16(c)) has been made use of in writing the solution (3.4.18(b)).

The constants A_1, B_1, A_2, B_2, C_{1n}, D_{1n}, C_{2n}, g_o, g_n, H_o and H_n are to be determined by substituting the appropriate quantities from Eqn. (3.4.18) in Eqns. (3.4.12)-(3.4.14) and (3.4.16).

As is clear from the above analysis, a knowledge of the constants g_o, H_o, g_n and H_n is sufficient to calculate the temperatures T_g and T_ω of the top cover and the saline water respectively. The heat flux from the water to the cover, involved in the evaporation process, is given by

$$q_{e\omega} = h_{eff}(T_\omega - T_g)$$

$$= h_{eff}[(H_o - g_o) + \text{Re} \sum_n (H_n - g_n) \exp(in\omega t)] \quad (3.4.19)$$

and the mass of water distilled per unit time per unit base area of the still is given by

$$\dot{m}_e = \frac{q_{e\omega}}{\mathcal{L}} \tag{3.4.20}$$

Note that $\mathcal{L}$ may be treated to be a constant in the temperature range of interest.

3.4.1(b). *The Mounted Still*

In this configuration (Fig. 3.2b), the bottom and the walls of the still are insulated and the still is mounted on a frame. The thoeretical analysis of this still follows on the same lines as that outlined for the ground-still. The energy balance at the glass cover and the interior base surface are still given by Eqs. (3.4.12)-(3.4.14). The energy balance at the outer surface of the insulation separating the still from the atmosphere is given by

$$-K_1 \left(\frac{\partial \theta}{\partial x}\right)_{x=L} = h_4 (\theta_{x=L} - T_a) \tag{3.4.21}$$

Here the wall thickness of the outer wooden casing used if any, is neglected. One may once again write down a periodic solution of the type

$$\theta(x,t) = [Ax + B + \mathrm{Re} \sum_n \{C_n \exp(-\beta_n x) + D_n \exp(\beta_n x)\} \times \exp(in\omega t)] \tag{3.4.22(a)}$$

$$T_g(t) = g_o + \mathrm{Re} \sum_n g_n \exp(in\omega t) \tag{3.4.22(b)}$$

$$T_\omega(t) = H_o + \mathrm{Re} \sum_n H_n \exp(in\omega t) \tag{3.4.22(c)}$$

where the constant coefficients are determined as shown earlier. The heat flux per unit area per unit time associated with the evaporation of water inside the still is given by

$$\mathcal{L} \dot{m}_e = q_{e\omega} = h_{eff} [(H_o - g_o) + \mathrm{Re} \sum_{n=1}^{\infty} (H_n - g_n) \exp(in\omega t)] \tag{3.4.23}$$

3.4.2. *Comparison of Experiment and Theory*

To test the validity of the periodic theory of the solar stills, a single basin solar still of base area 0.56 sq. meters was constructed by Nayak *et al.* (1980) using galvanized iron sheets of 24 SWG. A glass sheet, 3 mm thick, was used as the top cover. The vertical heights of the still were 29 cm and 15 cm with a slope of $10°$ along the breadth of the still. A V-channel, made of galvanized iron sheet, was fixed inside the still for collecting and draining distilled water. The interior of the still was painted black using blackboard paint. The entire structure was encased in a wooden box with a 5 cm thick layer of glass wool in between to act as an insulator.

The still was filled with water up to a depth of 10 cm and was left in the open facing south for 3 days so as to simulate the periodic conditions. The amount of distillate collected, solar intensity and the ambient temperature were recorded at hourly intervals. Temperatures of the glass cover and the water in the basin were monitored by means of copper-constantan thermocouples using a D.C. microvoltmeter; the other junctions of the thermocouples were immersed into an ice-bath. The water temperature was calculated as the average of several thermocouples fixed at different depths.

Single Basin Solar Still

The various parameters, necessary for performing numerical computations, of the still constructed were

h_1 = 16.076 W/m² °C L = 0.05 m
h_2 = 40.88 W/m² °C ρ_1 = 64.08 Kg/m³
h_{eff} = 8.555 W/m² °C C_1 = 670 J/Kg °C
h_3 = 137.373 W/m² °C (McAdams, 1954) K_1 = 0.04 W/m °C
h_4 = 22.71 W/m² °C (Duffie, 1974) τ_1 = 0.1
M_g = 5226 J/m² °C τ_2 = 0.0
$M_{\omega o}$ = 419000 J/m² °C τ_3 = 0.7

In the range of temperature variation observed in the experiment, the usual least square curve fitting procedure yielded the following linear relationship for the saturated vapor pressure, P, inside the still

$$P = 293.3 \; T - 84026.4 \; ,$$

where P is in N/m² and T in °K; data for the curve fitting was taken from steam tables (E. Schmidt, 1969).

Numerical calculations were performed for the still. Up to six harmonics were used in the Fourier expansion of the solar insolation and ambient temperature (Tables 3.3 and 3.4). The hourly production of distilled water per unit base area of the still is shown in Fig. 3.3; the experimental observations are represented by ⊙. The time variation of the glass and basin water temperatures is shown in Fig. 3.4. Here also ⊙ correspond to experimental measurements. In both instances the agreement between experiment and theory is very good. Thus, the retention of only six harmonics is found to be quite sufficient for predicting the performance of a solar still to a good degree of accuracy.

To illustrate the effect of insulation at the bottom of the still, the daily productivity of the still as given by

$$M_e = \int_0^{24 \times 60 \times 60} h_{eff} \frac{(T_\omega - T_g)}{L} \, dt$$

$$= \frac{h_{eff} \, a_o}{L \, U_L (h_1 + h_2)} \left[-\tau_1 U_b + \tau_2 h_2 + \tau_3 h_2 U_b \left(\frac{1}{h_4} + \frac{L}{K_1} \right) \right] \times$$

$$24 \times 60 \times 60 \; \frac{K_g}{m^2 day} \qquad (3.4.24)$$

is plotted with insulation thickness in Fig. 3.5. It may be seen that the productivity increases rapidly with increasing insulation thickness up to 4 cm and then it increases rather slowly.

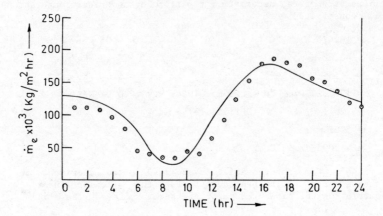

Fig. 3.3. Hourly variation of evaporated mass of water per unit basin area of mounted still.— Theoretical; ⊙⊙⊙ experimental points. (After Nayak *et al.*, 1980.)

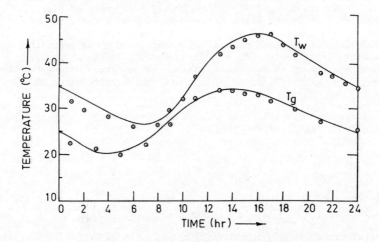

Fig. 3.4. Hourly variation of water and glass cover temperature of mounted still.— Theoretical; ⊙⊙⊙ experimental points. (After Nayak *et al.*, 1980.)

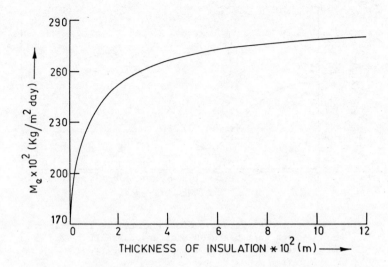

Fig. 3.5. Variation of average production per unit basin area of mounted still with thickness of insulation. (After Nayak et al., 1980.)

For the same solar intensity and ambient air temperature conditions, the daily productivity of the still, and the glass and water temperatures have been calculated for the ground-based solar still also using the following parameters:

K_1 = 0.04 W/m °C $\quad$ h_2 = 40.88 W/m² °C
L = 0.05 m $\quad$ h_3 = 137.373 W/m² °C
K_2 = 0.3463 W/m °C $\quad$ h_{eff} = 8.555 W/m² °C
α_1 = 6.2467 m^{-1} $\quad$ M_g = 5226 J/m² °C
α_2 = 11.8396 m^{-1} $\quad$ M_ω = 419000 J/m² °C
h_1 = 16.076 W/m² °C $\quad$ τ_1 = 0.1
τ_2 = 0.0 $\quad$ τ_3 = 0.7

The temporal variation of the productivity is shown in Fig. 3.6. Comparison of these results with that of the experimental still shows that the ground still is only marginally better than the mounted still.

Numerical calculations were also made to test the periodic theory of solar still against the experimental measurements of Cooper (1973b). The daily production of distilled water reported by Cooper (1973b) corresponding to the 4th March was 1.2 lb/ft² while the theoretical prediction turns out to be 1.349 lb/ft²; the parameters mentioned above were used in the calculations. The agreement between the two values could not be better because of the inadequacy of available data about the still used by Cooper (1973b).

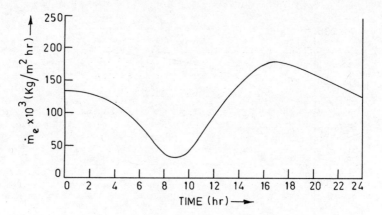

Fig. 3.6. Hourly variation of evaporated mass of water per unit basin area of ground still. (After Nayak et al., 1980.)

It is clear from these discussions that the performance of a single basin solar still can be predicted to a very good approximation by the model based on the periodic theory presented above. Some of the broad conclusions that can be drawn on the basis of this theory are as follows:

1. An inspection of the expression (Eqn. (3.4.19)) for the heat transfer rate, $q_{e\omega}$, associated with evaporative mass transfer, in the case of a ground still shows that the daily production of distilled water is independent of the insulation thickness below the base of the still. This is in agreement with the observations of Bloemer et al. (1965).

2. The theory is valid only for a still containing a large quantity of saline water; in this case the evaporative heat loss is much smaller than the heat content of saline water in the basin. This situation is represented by the asymptotic part of Fig. 3 of Bloemer et al. (1965) and Fig. 8 of Cooper (1969a). If however, a lower water depth is maintained, as desired for maximum productivity, the amount of distillate cannot be estimated by the present theory. But from the maintenance point of view a large water depth is preferred (Bloemer et al. 1965).

3. In the case of a still with a constant level arrangement for even low water depth, this model can very well be applied in estimating the production rate of the distillate.

4. The theory is based on the linearization of the heat and mass transfer relations. This, though a questionable approach, is justified because of the fact that the variations of temperatures of glass and water are small in a normal operation of the still. So the saturation vapor pressure can be assumed to have a linear dependence on temperature. Further, little error is committed in considering the heat transfer coefficients to be constant in this temperature range.

5. Not more than one harmonic had been used in earlier analyses of the solar still. It is impossible to reproduce solar intensity and ambient air temperature with one harmonic only. Hence one needs to consider more harmonics. As has been mentioned earlier six harmonics are found to give a good representation of the observed variation.

Single Basin Solar Still

6. Typical efficiency of the still is 30%, corresponding to a daily distillate of 2.4 Kg/m² day (on March 16, 1979 in Delhi).

3.5. EFFECT OF DYE ON THE PERFORMANCE OF A SOLAR STILL

In a conventional solar still discussed in the previous sections, most of the energy is absorbed at the base of the still which, therefore, becomes hotter than the water in the still. Although most of the heat is transferred to the water, a significant amount is also lost to the ground or the atmosphere by conduction through the base. Garg and Mann (1976) suggested a novel method of overcoming this difficulty, viz. the addition of a black dye to the saline water in the still. As a consequence, almost all of the solar energy entering the still gets absorbed by the water itself and the base does not get heated much (and consequently loses less heat). Thus, the water temperature in the presence of a dye is much higher than that in the absence of the dye; this leads to a marked improvement in the productivity of the still by addition of a dye.

The effect of a dye on the performance of a solar still has been systematically investigated by Rajvanshi and Hsieh (1979), by considering the base to be black and the water column to be a thermally stratified system. A detailed computer program was evolved to study the energy exchange between these layers and the base, good agreement between theory and experiment was reported.

In this section the periodic theory of stills, developed earlier, is used to analyse the performance of a solar still when a dye is present in the water; the results of the analysis are seen to be in good agreement with the experiment.

3.5.1. Numerical Calculations and Discussion

Two single basin stills were fabricated for the purpose, one to be used with plain water and the other contained water with the dye. The relevant parameters of these stills were:

Without dye		With dye	
ω	$= 7.2722 \times 10^{-5}$ s⁻¹	ω	$= 7.2722 \times 10^{5}$ s⁻¹
ρ_1	$= 64.08$ Kg/m³	ρ_1	$= 64.04$ Kg/m³
C_1	$= 670$ J/Kg °C	C_1	$= 670$ J/Kg °C
K_1	$= .04$ W/m °C	K_1	$= .04$ W/m °C
L	$= .05$ m	L	$= .05$ m
M_g	$= 5226$ J/m² °C	M_g	$= 5226$ J/m² °C
$M_{\omega o}$	$= 672634$ J/m² °C	$M_{\omega o}$	$= 672634$ J/m² °C
τ_1	$= 0.1$	τ_1	$= 0.1$
τ_2	$= 0.0$	τ_2	$= 0.8$
τ_3	$= 0.6$	τ_3	$= 0.06$
h_1	$= 22.52$ W/m² °C	h_1	$= 24.42$ W/m² °C
h_2	$= 50$ W/m² °C	h_2	$= 50$ W/m² °C
h_3	$= 135.05$ W/m² °C (Threlkeld, 1970)	h_3	$= 67.53$ W/m² °C (Threlkeld, 1970)

h_4 = 22.08 W/m² °C (Duffie, 1974)
α_1 = 6.2472 m⁻¹
h_{eff} = 14.01 W/m² °C

h_4 = 22.08 W/m² °C (Duffie, 1974)
α_1 = 6.2472 m⁻¹
h_{eff} = 15.51 W/m² °C

The difference in the values of τ_2 and τ_3 in the two cases is due to the presence of dye. Since the mean water temperature would be higher in the presence of the dye, the values of h_1, h_3 and h_{eff} are also different in the two cases. In all other respects the two stills are identical as may be seen from the above table.

The saturation vapor pressure as a function of temperature may be expressed by the following linear relation obtained by least square fitting of data in the temperature range (15°-65°C)

$$P = 420.69\ T - 1.22239 \times 10^5$$

where P is expressed in N/m² and T in °K. The Fourier coefficients corresponding to the periodic solar insolation and ambient temperature are given in Tables 3.5 and 3.6, respectively; the tables correspond to June 19, 1979 in Delhi.

TABLE 3.5. Fourier Coefficients of Solar Intensity Available on the Glass Surface

n	0	1	2	3	4	5	6
A_n (W/m²)	306.0554	457.165	157.995	18.917	22.051	18.089	7.6910
σ_n (degrees)		188.961	20.870	4.459	235.161	183.168	48.247

TABLE 3.6. Fourier Coefficients of Ambient Air Temperature

n	0	1	2	3	4	5	6
B_n (°C)	36.6958	0.666	1.168	0.045	0.300	0.298	0.157
ϕ_n (degrees)		230.105	351.466	77.879	155.915	299.941	25.516

The observed daily production of distilled water when different dyes are used is given in Table 3.7. Figure 3.7 shows the observed distilled water output as a function of the water depth in the still. The calculated hourly variation of the distillate per unit basin area, with and without the use of a dye, is shown in Fig. 3.8; the experimental points are shown by 'O'. The calculated values are seen to be in good agreement with the experimental observations.

TABLE 3.7. Effect of Dye on Daily Distillate

S. No.	dye used	Daily productivity for .1m depth/m^2		$\dfrac{\text{Output with dye}}{\text{Output without dye}}$	Total insolation Kj/m^2 day
		With dye (litre)	Without dye (litre)		
1.	Red	3.3037	3.066	1.08	2.477×10^4
2.	Violet	3.799	3.09	1.23	2.8713×10^4
3.	Black	3.862	3.066	1.26	2.477×10^4

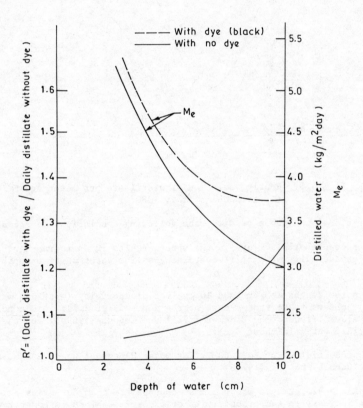

Fig. 3.7. Variation of the ratio of outputs with and without black dye (R) and distillate water with and without dye with water depth. (After Sodha *et al.*, 1980d.)

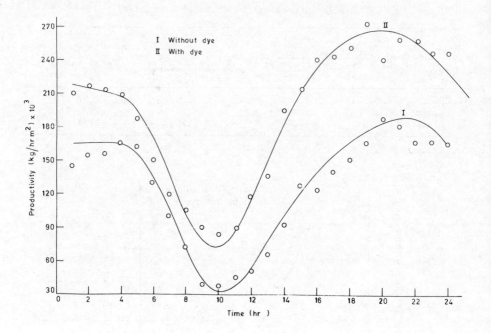

Fig. 3.8. Hourly variation of distillate per meter square. (After Sodha *et al.*, 1980d.)

On the basis of the above studies, the following conclusions may be drawn:

1. The presence of black dye in the water results in an increase of 26% in the productivity of a still containing saline water up to a depth of 10 cm.

2. For water depths larger than 10 cm, the productivity (with and without dye) appears to be independent of the depth (Fig. 3.7); this conclusion was also arrived at by Bloemer (1965) and Cooper (1969a), for the case of plain water (without dye).

3. The productivity is about 60%. For water depths of about 2 cms, as also concluded by Cooper (1973b).

4. Black and violet dyes are the most appropriate ones for bringing about this increase in the productivity of the still. This is understandable in terms of the spectral distribution of solar radiation.

5. For small depths of water the use of dye leads to only marginally better performance.

3.6. TRANSIENT ANALYSIS OF SOLAR STILL

3.6.1. *Background*

Morse and Read (1968) and Cooper (1969a) have amongst others developed graphical and numerical methods to predict the transient performance of a basin type solar still. Explicit expressions for the time dependent performance parameters have, however, not been obtained by them. In section 3.6.2 explicit expressions for the still parameters have been obtained, assuming steady state periodic conditions. Since the meteorological parameters are not strictly periodic over a period of days and the establishment of steady state periodic behaviour takes finite time (even for periodic meteorological parameters), Sodha *et al.* (1980a) have developed an analytical transient model for the behaviour of the still; simple expressions corresponding to periodic meteorological conditions (not periodic behaviour of the still) have also been given. The mass of water in the basin has been assumed to be constant, which implies the same approximations as in section 3.4.

3.6.2. *Analysis*

The energy balance for the glass cover and the water mass, may be expressed as,

$$M_g \frac{dT_g}{dt} = \tau_1 H_s + h_1(T_\omega - T_g) - h_2(T_g - T_a) \qquad (3.6.1)$$

and

$$M_{\omega o} \frac{dT_\omega}{dt} = \tau_2 H_s - h_1(T_\omega - T_g) + h_3(\theta_{basin} - T_\omega) , \qquad (3.6.2)$$

where the expressions for h's are the same as used in section 3.4.

In writing the approximate expression for h_{eff}, we have made the same assumptions as in the case of the periodic theory of solar still.

The energy balance for the basin liner, can be written as (Morse and Read, 1968).

$$\tau_3 H_s = h_3(\theta_{basin} - T_\omega) + h_b(\theta_{basin} - T_a) \qquad (3.6.3)$$

where,

$$\frac{1}{h_b} = \frac{L}{K_1} + \frac{1}{h_4}$$

Substituting the values of θ_{basin} from Eqn. (3.6.3) in Eqn. (3.6.2), one obtains

$$M_{\omega o} \frac{dT_\omega}{dt} = \tau_4' H_s - h_1(T_\omega - T_g) - U_b(T_\omega - T_a) \qquad (3.6.4)$$

where

$$\tau_4' = \tau_2 + U_b \left(\frac{L}{K_1} + \frac{1}{h_4} \right) \tau_3 ,$$

$$U_b = \left(\frac{1}{h_3} + \frac{1}{h_b} \right)^{-1} .$$

Equations (3.6.1) and (3.6.4) can be put in the following form,

$$\frac{dT_g}{dt} = a_1 T_\omega + b_1 T_g + f(t) \ , \qquad (3.6.5)$$

and

$$\frac{dT_\omega}{dt} = a_2 T_\omega + b_2 T_g + g(t) \qquad (3.6.6)$$

where,

$$a_1 = \frac{h_1}{M_g} \ , \quad b_1 = -\frac{(h_1 + h_2)}{M_g} \ , \quad f(t) = \frac{\tau_1 H_s + h_2 T_a}{M_g}$$

$$a_2 = -\frac{(h_1 + U_b)}{M_{\omega o}} \ , \quad b_2 = \frac{h_1}{M_{\omega o}} \ , \quad g(t) = \frac{\tau_4' H_s + U_b T_a}{M_{\omega o}}$$

Multiplying Eqn. (3.6.5) by α and adding it to (3.6.6), one obtains

$$\frac{d}{dt}(T_\omega + \alpha T_g) = T_\omega(a_2 + \alpha a_1) + T_g(b_2 + \alpha b_1) + s(t) \qquad (3.6.7)$$

where,

$$s(t) = g(t) + \alpha f(t)$$

Equation (3.6.7) can also be written as

$$\frac{d}{dt}(T_\omega + \alpha T_g) = c(T_\omega + \alpha T_g) + s(t) \qquad (3.6.8)$$

where

$$c = a_2 + \alpha a_1 \ ,$$

and

$$c\alpha = b_2 + \alpha b_1$$

which gives

$$\alpha_\pm = \frac{-(a_2 - b_1) \pm \sqrt{(a_2 - b_1)^2 + 4a_1 b_2}}{2a_1} \ .$$

Thus Eqn. (3.6.8) can be written as

$$\frac{d}{dt}(T_\omega + \alpha_\pm T_g) = C_\pm(T_\omega + \alpha_\pm T_g) + S_\pm(t) \qquad (3.6.9)$$

where

$$C_\pm = a_2 + \alpha_\pm a_1$$

and

$$S_\pm(t) = g(t) + \alpha_\pm f(t)$$

The solution of Eqn. (3.6.9) is of the form

$$T_\omega + \alpha_\pm T_g = \exp(C_\pm t) \int S_\pm(t) \exp(-C_\pm t) dt + A_\pm \exp(C_\pm t) \ , \qquad (3.6.10)$$

Single Basin Solar Still   39

where $A_{\pm}$ are the constants to be determined with the help of boundary conditions, viz.

$$T_\omega(t=0) = T_{o_1} \text{ and } T_g(t=0) = T_{o_2}$$

To obtain the solution in a closed form, we may express the insolation H_s and atmospheric temperature T_A as a Fourier series in time (Eqn. (3.4.1)).

From Eqn. (3.6.10), the boundary conditions and the series expansion for $H_s(t)$ and $T_A(t)$, one obtains,

$$T_\omega(t) = \frac{1}{\alpha_- - \alpha_+} \left[(\alpha_- P_3 - \alpha_+ P_4) + \sum_n (\alpha_- P_{1n} - \alpha_+ P_{2n}) \exp(in\omega t) \right.$$
$$\left. + A_+ \alpha_- \exp(c_+ t) - A_- \alpha_+ \exp(c_- t) \right] , \qquad (3.6.11)$$

and

$$T_g(t) = \frac{1}{\alpha_+ - \alpha_-} \left[(P_3 - P_4) + \sum_n (P_{1n} - P_{2n}) \exp(in\omega t) \right.$$
$$\left. + A_+ \exp(c_+ t) - A_- \exp(c_- t) \right] \qquad (3.6.12)$$

where P_3, P_4, P_{1n}, and P_{2n} can be obtained from Eqn. (3.6.10). Hence the heat flux corresponding to evaporation from water can be written as

$$q_e = \dot{m}_e \mathcal{L} = h_{eff}(T_\omega - T_g) \qquad (3.6.13)$$

The amount of distillate per unit area per hour is given by

$$\dot{m}_e = \frac{h_{eff}(T_\omega - T_g)}{\mathcal{L}} \times 60 \times 60 \; \frac{Kg}{m^2 hr} \qquad (3.6.14)$$

This analysis is found to be in good agreement with experiments (Sodha *et al.* 1980).

3.7. NOCTURNAL PRODUCTION OF SOLAR STILL

The term 'nocturnal production' is used in reference to the operation of a solar still in the absence of sunlight as the direct source of energy. In such a situation either the solar energy stored during sunshine hours is used during night or the use of waste heat available from various industries and power plants (say in the form of discharged hot water) can enhance the productivity appreciably. If the stills are fed with the available hot water throughout the day, the still productivity can further be augmented.

The earliest work on nocturnal operation of the solar still was reported by Grune *et al.* (1962); they used a deep basin solar still, operated with, forced circulation of both water and humid air by pump and fans with continuous production throughout the 24 hour period. It was concluded that the use of warm water increases the daily yield of a solar still significantly. The analysis and performance of the different modes of nocturnal operation of solar stills are discussed below in detail.

3.7.1. Tubular Solar Still*

Tleimet and Howe (1966) designed a special still for nocturnal operation. In order to exploit the available energy to the maximum extent during non-sunshine hours, they simply used a glass tube, 1.17m long with internal diameter equal to 0.12m and wall thickness of 3 mm, as the main chamber (see Fig. 3.9). The ends of the tube were closed by means of wooden planks; rubber gaskets were used to make the system airtight. A shallow metal tray, 0.1m wide, 1.12m long and 1.27cm deep was fitted inside the tube and supported at the wooden end plates. The tray was painted black and it formed the basin of the still. Saline water was allowed to flow into the still from a separate constant waterhead tank system; the extent of water head could be altered by changing the location of the overflow hole in the tank. Thus, the rate at which the saline water was fed into the still could also be controlled.

Saline water in the tray got heated due to the absorption of diffuse radiation from almost all sides of the still. The water vapor emitted then condenses on the cooler wall of the tube and flows down to the bottom of the tube from where it is piped out. The brine that accumulates in the tray is discharged to the drain.

The distillate was collected in a recording-type rain gauge and measured at the end of the test run. The chart on the rain gauge showed the production rate to be constant during the whole run.

The production of distillate as a function of the temperature difference between the inlet water and the minimum ambient air is given by the empirical equation

$$M_e = 0.1323 \ W^{0.3} \ (T_{in} - T_a) - 1060$$

where

M_e = production of distillate Kg/m^2 day
W = brine flow rate, Kg/m^2 hr
$(T_{in} - T_a)$ = temperature difference between inlet water and ambient air, $^\circ C$.

3.7.2. Single Basin Solar Still**

The schematic set-up of a nocturnally producing ground based solar still is given in Fig. 3.1b. The heat and mass transfer relationships governing the nocturnal output of a solar still are essentially the same as those discussed in section 2.

Assuming steady state conditions, the energy balance on the cover results in

$$q_{e\omega} + q_{r\omega} + q_{c\omega} = q_{ca} + q_{ra} \qquad (3.7.1)$$

where,

$q_{c\omega}$, $q_{e\omega}$, $q_{r\omega}$, q_{ra} and q_{ca} are same as Eqns (2.1.7), (2.1.19), (2.1.20), (2.2.1) and (2.2.2) respectively. The energy balance on the brine-mass gives:

---

*This is discussed here for the sake of completeness in the discussion of nocturnal production.

**After Malik and Tran (1973).

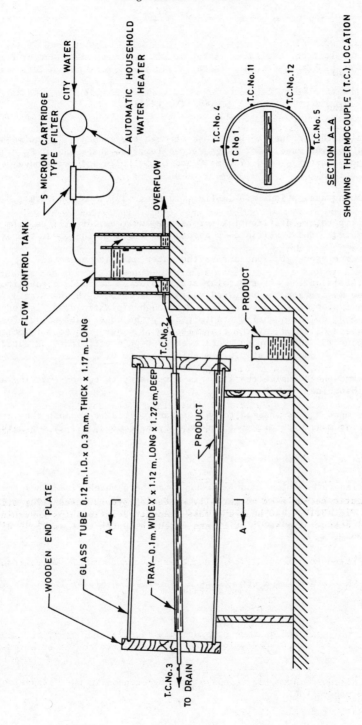

Fig. 3.9. Arrangement of tubular glass still for nocturnal production. (After Tleimat and Howe, 1966.)

$$q_{e\omega} + q_{r\omega} + q_{c\omega} + q_{ins} = - M_{\omega o} \frac{dT_\omega}{dt} \qquad (3.7.2)$$

where,

$$q_{e\omega} = \dot{m}_e \mathcal{L}$$

$$q_{ins} = \frac{S_s K_2}{L_{s\omega}} (T_\omega - T_a) \qquad \text{(Keller, 1928)}$$

To determine the conduction heat loss through the ground q_{ins} the method of Keller (1928) was used. For large rectangles, the shape factor $S_s = 3.73$ is used. The value of $K_2 = 0.87$ W/m °C and $L_{s\omega} = 1.22$m and 2.44m have been used in calculations.

The following assumptions are implicit in Eqns (3.7.1) and (3.7.2):

1. The temperature differential across the cover and distillate film is neglected. The distillate is assumed to leave at the cover temperature.

2. The temperature gradient across the water mass is neglected.

3. Water film on the plastic solar still cover is assumed to constitute an opaque medium for the infrared radiation emitted by brine mass. The transmission of infrared radiation through a plastic cover is very dependent on the quantity transmitted by the condensate film. For a film thickness of 0.002 cm, the transmission of water, for normal incidence, is about 25% (based on an average absorption coefficient for water of 690 cm^{-1}). However, at 0.01 cm this drops to 0.1%.

4. The atmosphere inside the still is considered to be non-emitting and non-absorbing.

5. Brine mass $M_{\omega o}$ is assumed to be constant. This is because the variation in brine mass with time for moderate brine depths, ℓ_ω, is negligibly small.

3.7.2(a). *A Simplified Mathematical Model*

A simplified mathematical model is suggested here which linearizes the heat and mass transfer relationships in a still. Closed form solutions are obtained for predicting the nocturnal output (expressed as a dimensionless number $m_\omega/M_{\omega o}$) as functions of initial brine temperature $T_{\omega o}$ and drop in brine temperature $(T_{\omega o} - T_\omega)$.

From Eqn. (3.7.1)

$$h_{eff}(T_\omega - T_g) + h_{r\omega}(T_\omega - T_g) + h_{c\omega}(T_\omega - T_g) = h_{ca}(T_g - T_a) + h_{ra}(T_g - T_a)$$

or

$$h_1(T_\omega - T_g) = h_2(T_g - T_a) \qquad (3.7.3)$$

where

Single Basin Solar Still

$$h_{eff} = \frac{q_{e\omega}}{(T_\omega - T_g)},$$

$$h_{c\omega} = \frac{q_{c\omega}}{(T_\omega - T_g)},$$

$$h_{ra} = \frac{q_{ra}}{(T_g - T_a)},$$

$$h_{r\omega} = \frac{q_{r\omega}}{(T_\omega - T_g)},$$

$$h_{ca} = \frac{q_{ca}}{(T_g - T_a)},$$

$$h_1 = h_{eff} + h_{r\omega} + h_{c\omega},$$

$$h_2 = h_{ca} + h_{ra}$$

The energy balance on the water mass results in:

$$h_1(T_\omega - T_g) + h_b(T_\omega - T_a) = -C_{p\omega}\frac{d}{dt}(M_R T_\omega)$$

$$= -C_{p\omega}\left[M_R \frac{dT_\omega}{dt} + T_\omega \frac{dM_R}{dt}\right] \quad (3.7.4)$$

When T_g is eliminated from Eqns (3.7.3) and (3.7.4), one obtains

$$(T_\omega - T_a)\left[\frac{h_1 h_2}{h_1 + h_2} + h_b\right] = -C_{p\omega}\left[M_R \frac{dT_\omega}{dt} + T_\omega \frac{dM_R}{dt}\right] \quad (3.7.5)$$

The water mass balance can be expressed as

$$\frac{h_{eff}}{\mathcal{L}}(T_\omega - T_g) = -\frac{dM_R}{dt}$$

(assuming $\mathcal{L} = h_{g\omega} - h_{fC}$).

The cover temperature may be eliminated by using the following relation, derived from Eqn. (3.7.3)

$$(T_\omega - T_g) = \left[\frac{h_2}{h_1 + h_2}\right](T_\omega - T_a).$$

Hence

$$\frac{h_{eff}}{\mathcal{L}}\left(\frac{h_2}{h_1 + h_2}\right)(T_\omega - T_a) = -\frac{dM_R}{dt} \quad (3.7.6)$$

Dividing Eqn. (3.7.5) by Eqn. (3.7.6) yields

$$\frac{dT_\omega}{(dM_R/M_R)} + T_\omega = \frac{h_1 h_2 + h_b(h_1 + h_2)}{C_{p\omega} h_{eff} h_2} \cdot \mathcal{L} = \lambda \qquad (3.7.7)$$

where λ is a constant.

The solution to Eqn. (3.7.7) subject to the boundary conditions

$$\text{at } t = 0, \; T_\omega = T_{\omega o} \; ; \; M_R = M_{\omega o}$$

is

$$\frac{m_\omega}{M_{\omega o}} = \frac{M_{\omega o} - M_R}{M_{\omega o}} = \frac{T_{\omega o} - T_\omega}{\lambda - T_\omega} \qquad (3.7.8)$$

It is seen from Eqn. (3.7.8) that, if λ is a constant, the distillate depends primarily on the initial water mass $M_{\omega o}$ (and hence the water depth ℓ_ω), initial water temperature $T_{\omega o}$ and drop in water temperature $(T_{\omega o} - T_\omega)$.

The following assumptions are implicit in the analysis cited above:

1. h_{eff}, $h_{c\omega}$, $h_{r\omega}$ and, hence, their sum h_1 are constant.
2. h_{ca}, h_{ra} and, hence, their sum h_2 are constant.
3. $\mathcal{L}$, h_b, $C_{p\omega}$ and λ are constant.

By choosing the proper value of λ for a given initial brine temperature, Eqn. (3.7.8) can be used for determining the distillate m_ω as a function of brine depth ℓ_ω initial brine temperature $T_{\omega o}$ and drop in temperature $(T_{\omega o} - T_\omega)$. The resulting expressions which predict the nocturnal output of a still for four different initial brine temperatures are as follows:

Initial brine temperatures °C	Equation for distillate
$T_{\omega o} = 65.5$	$m_\omega / M_{\omega_o} = (65.5 - T_\omega)/(1153 - T_\omega)$
$T_{\omega o} = 60.0$	$m_\omega / M_{\omega_o} = (60.0 - T_\omega)/(1230 - T_\omega)$
$T_{\omega o} = 54.5$	$m_\omega / M_{\omega_o} = (54.5 - T_\omega)/(1242 - T_\omega)$
$T_{\omega o} = 49.0$	$m_\omega / M_{\omega_o} = (49.0 - T_\omega)/(1346 - T_\omega)$

3.7.2(b). Experimental Investigation

During the summer of 1969 an experimental still of dimensions 3.66×1.22m, was built by Malik and Tran (1973), which was covered with an inflated plastic cover. The still was located on the roof of the Machinery Hall at the MacDonald campus of McGill University of Quebec, Canada (latitude 45° 26', longitude 73° 57').

The still was constructed of 0.025m cedar planks. The base was insulated with paper rolled in a ball. No insulation was used for the sides and ends other than an extra plank of wood. This also served to strengthen the sides. The whole still was painted with aluminium paint which helped to reduce heat losses. In the basin liner, a 30 mil (0.76m) thick, black, butyl-rubber sheet was laid directly over the hardboard.

Malik and Tran (1973) found the analysis presented in section 3.7.2(a) to be in good agreement with experiment and concluded that:

1. Lower values of relative humidity (γ) (and hence T_{sky}) and T_a, and higher value of ϑ, will result in quicker cooling of the brine mass. Similarly, higher values of γ, and lower values of ϑ will retard the cooling of the brine mass. However, the production obtained in each case in cooling the water mass from $T_{\omega o}$ to T_ω is the same; there is merely a saving in time in the first case.

2. The dimensionless ratio $m_\omega/M_{\omega o}$ is a function only of the initial brine temperature and the drop in water temperature. It is independent of water depth, ambient air temperature, relative humidity, least width dimension and time.

3. In the case where all parameters – except water depth – are constant, the brine mass with a smaller depth cools more rapidly than the one with larger depth. However, the ratio $m_\omega/M_{\omega o}$ remains the same as long as T_ω and $(T_{\omega o}-T_\omega)$ are the same. It means that if the brine depth is reduced, a reduction in the time required for obtaining the same value of $m_\omega/M_{\omega o}$ results.

4. Since $q' = M_{\omega o} C_{p\omega}(T_{\omega o}-T_\omega)$, it is obvious that, for a fixed heat input q', decreasing water depth will result in a higher initial saline water temperature. The amount of distillate obtained by letting the water cool in the still from initial water temperature $T_{\omega o}$ to a final temperature $T_F = T_i$ is given by

$$m_\omega = \frac{q'}{C_{p\omega}(\lambda - T_i)}$$

It is seen from this equation that for a given amount of heat, nocturnal output decreases as the brine depth is increased. This is because an increase in the brine depth would reduce initial brine temperature which in turn would raise the value of λ. Brine depth is, therefore, an important parameter and should be kept as low as possible. In other words, one should aim at obtaining the highest possible initial brine temperature with the amount of heat available.

5. The highest temperature permissible without scaling should be employed. Its highest value may be taken as 99°C.

6. Significance of initial water temperature:
Suppose that a certain brine mass has been heated to an initial brine temperature $T_{\omega o}$. Consider the effect of heating it further to a temperature $(T_{\omega o} + \Delta T_\omega)$. The amount of heat needed to heat water to the new temperature is:

$$q' = M_{\omega o} C_{p\omega} \Delta T_\omega \qquad (3.7.9)$$

The amount of distillate obtained by the time the brine mass has cooled to temperature $T_{\omega o}$ is

$$\frac{m_\omega}{M_{\omega o}} = \frac{\Delta T_\omega}{\lambda - T_{\omega o}} \qquad (3.7.10)$$

Dividing Eqn. (3.7.10) by Eqn. (3.7.9) one obtains

$$\frac{m_\omega}{q'} = \frac{1}{C_{p\omega}} \left[\frac{1}{\lambda - T_{\omega o}} \right]$$

Since an increase in initial brine temperature results in monotonic decrease in the value of the factor $(\lambda - T_{\omega o})$ it is clear from Eqn. (3.7.8) that the amount of distillate per unit of energy m_ω/q would increase with a rise in $T_{\omega o}$.

7. Even if waste heat is available, other sources of energy must be considered as supplementary sources. The amount of energy needed to heat a given brine mass in a solar still from, say, 43-49°C is the same as that required for heating it from 65-71°C. However, the distillate obtained when the brine mass cools from 71-65°C is more than that obtained by cooling it from 49-43°C. Thus, supplementary energy becomes more attractive as the amount of available waste heat increases.

3.7.2(c). *Rigorous Analytical Model**

An analysis of the operation of a single basin solar still in which waste hot/warm water is used, is presented here; linearized Dunkle's relations for heat and mass transfer processes have been used. Two modes of still operation have been considered:

(i) flowing waste hot water at constant rate through the solar still throughout the 24 hours,
(ii) feeding waste hot water once in a day only.

The analytical model developed earlier (section 3.6) is applicable to the operation mode (ii) with appropriate value of initial water temperature.

For operation in mode (i) of the still the analytical expressions for the hourly yield, basin water temperature, glass temperature and outlet water temperature may be obtained as follows:

The energy balance equation for the water moving in the basin along y-direction is

$$\left(\rho_w b d C_{pw} \frac{\partial T_w}{\partial t} + \dot{m}_w C_{pw} \frac{\partial T_w}{\partial y} \right) \cdot dy$$

$$= \left[\tau \cdot H_s - h_1(T_w - T_g) - U_b(T_w - T_a) \right] b \cdot dy \qquad (3.7.11)$$

where, U_b - the overall bottom loss coefficient and τ are defined as

---

*After Sodha *et al.* (1981b).

Single Basin Solar Still

$$U_b = \left(\frac{1}{h_3} + \frac{L}{K_1} + \frac{1}{h_4}\right)^{-1},$$

$$\tau = \left(\tau_2 + \frac{h_3 \tau_3}{h_3 + h_b}\right)$$

The energy balance for the glass cover of the still may be written as in Eqn. (3.6.1).

In view of periodic H_s and T_a, water and glass temperatures may also be assumed to be periodic i.e.,

$$T_w(y,t) = T'_{wo}(y) + \sum_{n=1}^{\infty} T'_{wn}(y) \exp(in\omega t)$$

and

$$T_g(y,t) = T_{go}(y) + \sum_{n=1}^{\infty} T_{gn}(y) \exp(in\omega t),$$

where the constants T'_{wo}, T_{go}, T'_{wn} and T_{gn} are to be determined with appropriate boundary conditions. The initial condition at y=0 is

$$T_w = T_{wo} \quad \text{for all values of t,}$$

where

$$T_{wo} = T_{oo} + \sum_{n=1}^{\infty} t_{on} \exp(in\omega t)$$

and

$$t_{on} = T_{on} \exp(-i\sigma'_n) \;;$$

T_{on} and σ'_n may be evaluated by Fourier analysis of the time dependence of inlet water temperature.

Thus for time independent part of T_g and T_w we obtain

$$T_{go} = \frac{1}{(h_1 + h_2)} \left[\tau_1 a_o + h_1 T'_{wo} + h_2 b_o\right],$$

and

$$\frac{dT'_{wo}}{T_1 - T'_{wo}} = \frac{bH}{\dot{m}_w C_w} \cdot dy \qquad (3.7.12)$$

where

$$T_1 = \frac{1}{H} \left[\left(\tau_4 + \frac{h_1 \tau_1}{h_1 + h_2}\right) a_o + \left(U_b + \frac{h_1 h_2}{h_1 + h_2}\right) b_o\right],$$

and

$$H = \left(U_b + \frac{h_1 h_2}{h_1 + h_2}\right)$$

Integrating (3.7.12) and using the initial condition $T'_{wo} = T_{wo}$ at $y=0$ one obtains,

$$T'_{wo} = T_1 + (T_{oo} - T_1) \exp(-\alpha y)$$

where

$$\alpha = \left(\frac{bH}{\dot{m}_w C_w} \right) .$$

Similarly for the time dependent part one can obtain

$$T_{gn} = \left[\frac{h_1 T'_{wo} + \tau_1 a_n + h_2 b_n}{h_1 + h_2 + in\omega M_g} \right] ,$$

and

$$T'_{wn} = T_{1n} + (t_{on} - T_{1n}) \exp(-\alpha' y) ,$$

where

$$T_{1n} = \frac{1}{H_n} \left[\left\{ \tau_4 + \frac{h_1 \tau_1}{(h_1 + h_2 + in\omega M_g)} \right\} a_n \right.$$
$$\left. + \left\{ U_b + \frac{h_1 h_2}{(h_1 + h_2 + in\omega M_g)} \right\} b_n \right] ,$$

$$H_n = \left[\frac{in\omega M_w}{b} + h_1 + U_b - \frac{h_1^2}{h_1 + h_2 + in\omega M_g} \right]$$

and

$$M_w = \rho_w d b C_{pw}, \quad \alpha' = \left(\frac{bH_n}{\dot{m}_w C_{pw}} \right) .$$

The rate of evaporative heat flux is

$$Q_e(y,t) = h_e \left[T_w(y) - T_g(y) \right]$$

$$= h_e \left[(T'_{wo} - T_{go}) + \sum_{n=1}^{\infty} (T'_{wn} - T_{gn}) \exp(in\omega t) \right] \quad (3.7.13)$$

Space average of $Q_e(y,t)$ can be obtained by integrating the above equation w.r.t.y. Hence

$$Q_e(t) = \frac{1}{L'} \int_0^{L'} Q_e(y,t) \, dy$$

$$= h_{eff} \left[(N_1 \bar{T}'_{wo} + N_2) + \sum_{n=1}^{\infty} (M_1 \bar{T}'_{wn} + M_2) \exp(in\omega t) \right]$$

where

$$N_1 = \left(\frac{h_2}{h_1 + h_2}\right),$$

$$N_2 = -\left[\frac{\tau_1 a_o + h_2 b_o}{h_1 + h_2}\right],$$

$$\bar{T}_{wo} = \left[T_1 + (T_{oo} - T_1) \cdot \frac{1 - \exp(-\alpha L')}{\alpha \cdot L'}\right],$$

$$M_1 = \left[\frac{h_2 + in\omega M_g}{h_1 + h_2 + in\omega M_g}\right],$$

$$M_2 = -\left[\frac{\tau_1 a_n + h_2 b_n}{h_1 + h_2 + in\omega M_g}\right]$$

$$\bar{T}_{wn} = \left[T_{1n} + (T_{on} - T_{1n}) \cdot \frac{1 - \exp(-\alpha'L')}{\alpha'L}\right]$$

The outlet water temperature ($y = L'$) may be obtained with the help of above analysis as

$$T_f(t) = T_w(y,t)\big|_{y=L'}$$

$$= T_1 + (T_{oo} - T_1) \exp(-\alpha L')$$

$$+ \sum_{n=1}^{\infty} \{T_{1n} + (t_{on} - T_{1n}) \exp(-\alpha'L')\} \exp(in\omega t) \cdot$$

The mean hourly yield of the still is given by

$$\dot{m}_e = \frac{Q_e(t)}{\mathcal{L}} \times 60 \times 60 \quad Kg/m^2 \cdot hr$$

3.7.2(d). *Numerical Results and Discussion*

To appreciate the analytical results numerical calculations corresponding to two typical days (June 19, 1979 and March 9, 1979) for Delhi were carried out for the two indicated modes of operation of still. The relevant parameters used were

(i) $\tau_1 = 0.1$ $h_4 = 6.27 \; W/m^2 \; °C$

$\tau_2 = 0.0$ $U_b = 0.7686 \; W/m^2 \; °C$

$\tau_3 = 0.6$ $b = 1.00 m$

$L = 0.05 m$

50   Solar Distillation

(ii) Constant Flow rate:

Summer

h_1 = 22.52 W/m² °C
h_2 = 50.0 W/m² °C
h_3 = 135.05 W/m² °C
h_{eff} = 14.01 W/m² °C
R_1 = 420.69 N/m² °C
R_2 = -7327.53 N/m²
T_{oo} = 40, 45, 50, 55°C
L' = 1-40m
U_o = 0.0005-0.010 m/sec
d = 0.01-0.15m
T_{on} = B_n, $\sigma'_n = \phi_n$

Winter

h_1 = 16.76 W/m² °C
h_2 = 40.88 W/m² °C
h_3 = 135.05 W/m² °C
h_{eff} = 8.55 W/m² °C
R_1 = 293.3 N/m² °C
R_2 = -39 11.505 N/m²
T_{oo} = 15, 30, 40°C

(iii) Once a day feed:

Summer

T_{go} = 37.5°C
T_{01} = 45. 55, 65°C
d = 0.15m

Winter

T_{go} = 12.5°C
T_{01} = 20, 30, 40°C
d = 0.15m

From the numerical analysis the following conclusions are drawn.

(i) Average daily yield increases with increasing inlet water temperature.
(ii) When $T_{wo}' < T_1$ or $T_{wo}' > T_1$ the conclusions for maximum productivity arrived at are,

   (a) for $T_{oo} > T_1$, one should have small L', and large d and U_o.
   (b) for $T_{oo} < T_1$, one should have large L', and small d and U_o.

(iii) For $T_{wo}' > T_1$, the yield is larger in mode 1, while for $T_{wo}' < T_1$, the yield is larger in mode 2.
(iv) The analysis should also find its application in the concept of scale prevention by constant flow of water.

3.8. PARAMETRIC STUDIES

3.8.1. Experience with Plastic Covers

The acceptance of the solar still on a large scale depends very much on its cost and the ease with which it can be transported, installed and repaired. It may be noted that the transparent cover is the single most expensive item of a still, which may also have to be replaced often. Thus, the use of transparent plastic sheets in place of glass cover would substantially reduce the overall cost of the still besides the additional advantages such as flexibility in design and elimination of airtight glass to container joints. Keeping these points in view, several small solar stills have been designed suitable for short term use (particularly in the developing countries (Lof, 1966)). In all these designs plastic sheets are used as the transparent covers and either concrete or black plastic film is employed for the basin of the still.

Single Basin Solar Still

For convenience these systems may be classified into three general types:

1. Inflated or wire-supported all-plastic tubular condensing surface with enclosed saline water tray (Bloemer et al., 1964).

2. Conventional horizontal basin with separate plastic covers supported by frames (Daniels, 1965).

3. Circular basin with vertical or inverted plastic conical cover (Howe, 1964).

Unfortunately, none of these systems have been subjected to rigorous field trials and hence their reliability in long term operation is questionable. Despite their low cost per unit still and the general adaptability in regions hampered by severe weather conditions, the use of plastics has not always yielded satisfactory results. Some of the common problems encountered are the following:

1. Fragility and short service life of plastic sheets.

2. Leakage of water vapor and the condensate.

3. Over-heating, and hence melting, of the plastic bottom of the still due to the development of dry spots in course of time. In the extreme case the black polyethylene sheets used as the basin liner may get heated beyond its melting point.

4. The plastic cover surface does not get wetted and this leads to reduced transmission of incoming solar energy and also to dripping of distilled water back into the brine liquid.

5. Susceptibility to damage by wind and other elements of nature.

6. Occasional unforeseen mixing of brine and distilled water in some of the designs.

Interest in the use of plastic films was revived in the spring of 1963, when a circular still, 2.44m in diameter, was designed and tested by Tleimat and Howe (1967). The first unit of this design, shown in Fig. 3.10, was installed early in the summer of 1963. The frame was made of concrete circular segments, and the cover was made from a cellulose-acetate film, .02m thick, which was shaped into a 3.05m diameter cone with an apex angle of 140 degrees. Nylon cords, emanating from the centre support and fastened at the edges, served as supports for the cover. The edges of the cover were secured by the weight of the circular segments placed on top of it. The water tray, made of black polyethylene film of thickness .015 cm, was held in a groove in the lower concrete segments to form the distillate trough. The first cover lasted about six months and failed because wind-fluttering produced cracks at the edges under the concrete segments leading to leakage of vapor. Also, the nylon cords started to rot and gradually broke, causing uneven loading on the cover.

Rain falling on the cover was collected around the periphery, and drained into the product trough through holes in the cover, thereby washing dust and sand into the distillate trough and clogging the outlet. The rain, collected at the periphery, loaded the cover to such an extent as to pull the edge from under the concrete segments, permitting the wind to catch the free edge and tear a large portion of the cover.

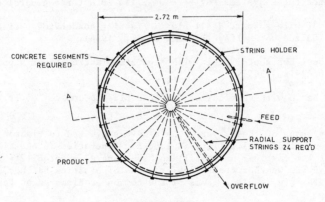
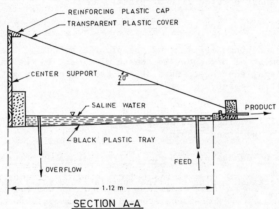

Fig. 3.10. Conical plastic covered circular still.
(After Tleimat and Howe, 1967.)

Inspection of the polyethylene water liner showed deterioration along the edges which separated the water tray and the product trough. The polyethylene sheet was replaced by black reinforced griffolyn fabric film.

By this time a technique was developed to render the Tedlar wettable by mechanically roughening the surface. The treatment was accomplished by scratching the surface with very fine grade sandpaper, using water as a lubricant. A new wettable Tedlar cover was made and installed; this lasted for approximately one year. The repeated flexing of the Tedlar cover under the concrete segments, due to dynamic wind-loading, caused the material to fatigue and crystallize so that it became brittle and failed along the edges.

One of the basic questions that arises in the use of plastic films for solar stills is its relative effectiveness compared to glass, under identical conditions of construction and operation. To find an answer to this question two small tilted-tray type, identical stills were constructed by Tleimat and Howe (1969) and put into operation in March 1964. The first was covered with window glass (3mm thick) and the second one was covered with a type-40 clear

Tedlar (.0005m thick, mechanically treated to produce a wettable surface). Extreme care was taken to ensure that the exposed areas to solar radiation of both stills were identical in size.

Examination of the data during the first month of operation indicated nearly equal productivity of both stills, after which the production of the plastic-covered still started to decline. Data were collected for about two years. The quantity of water produced by the two stills was averaged over the 2-year period. The total production of the Tedlar-covered still was about 82% of the total production of the glass-covered still for the same period. During this period, three small holes, possibly due to the weakening of the Tedlar film from the mechanical treatment, were observed on the plastic cover and were patched up with small strips of black vinyl adhesive tape to ensure vapor tightness. At the end of the test period, the Tedlar cover was removed and examined. Visual inspection of the cover showed no signs of further deterioration or failure from the dynamic wind-loading.

Thus, on the whole it appears that plastic sheets may not be the right choice for meeting the fresh water requirements of a community or family on a long term scale mainly because of its low capacity and doubtful reliability. However, they are best used in cases of emergency, war and in general for short time applications.

3.8.2. *Other Materials*

Apart from plastics, one has so many materials to choose for use in the fabrication of solar stills. The deciding factors are the cost and constraints peculiar to any particular location. Achilov *et al.* (1973a) have conducted extensive studies on the use of several materials such as concrete, sand concrete, ferroconcrete, foamed plastic, various types of sheet, plate and window glass, zinc coated and rubber piping, various types of steel, duralumin, various types of thermal insulators, asbestos cement, reinforced cement, dyes, lacquers and other cementing compounds in the fabrication of a still. An important outcome of these studies has been that, if the still is to be installed at any location permanently it is best to construct the basin of the still using sand concrete or water proofed concrete; Baum *et al.* (1970) also arrived at the same conclusion.

In the case of factory-manufactured system, prefabricated ferroconcrete is more convenient. The extensive use of prefabricated ferroconcrete offers several important advantages, including the mechanization of constructional work, possibility of construction work throughout the year, reduction in fabrication time, improvement in quality and reduction in cost. Moreover, there is a substantial economy in the use of timber because of the reduction in the use of casings and winter enclosures. Thus, if the cost of timber in the case of monolithic ferroconcrete still amounts to 7.5-12% of the total cost of the construction, in the case of prefabricated ferroconcrete this can be reduced to 1.0-2.5% and thus the cost of installation as a whole is reduced by 9%. Prefabricated ferroconcrete ensures a reduction in the consumption of concrete by 15%, a reduction in capital investment by 15-20%, and a reduction of 40% in the labor. If the manufacturing process be continued throughout the year, a 10-15% increase in the production rate can be achieved. Above all, prefabricated ferroconcrete still basins are transportable, convenient to use and independent of local conditions.

According to Baum *et al.* (1970) the type MP5-XII-2 insulating paste has been found to give the best results when used as a sealing compound. It exhibits good adherence to iron, various concrete types, wood, glass etc., is noted for its marked elasticity, undergoes no cracking and does not peel off under the effect of heat or moisture. Superior hydraulic insulation may be attained by placing a film between the reinforced concrete base of the still and the cement binder. Since this film initially prevents the seepage of the water to be demineralized, while at subsequent stages of solar still operation pores in the concrete and cement binder will be filled with a salt precipitate which provides a reliable hermetic seal of the solar still bottom.

3.8.3. *Effect of Meteorological and Still Parameters*

3.8.3(a). *Effect of Wind Velocity*

Cooper (1969a) concluded that for average wind velocities from 0 to 2.15 m/sec the output gets increased by 11.5%, while from 2.15 m/sec to 8.81 m/sec the increase was only 1.5%, indicating the decreasing influence of wind at higher velocities on distillation rate, as has also been concluded by Morse and Read (1968). A suggested fall in productivity with increasing wind velocity by Lof *et al.* (1961b) is not consistent with the present understanding of the heat and mass transfer modes.

Soliman (1972) has studied the effect of wind velocity on output in detail, considering all heat and mass transfer modes. He has concluded that at high water temperature, the increase in the difference between water and cover temperatures by increasing wind speed causes an increase in the rate of evaporation.

3.8.3(b). *Effect of Ambient Temperature*

Morse and Read (1968) have shown that an ambient temperature change from 26.7°C to 37.8°C causes an 11% increase in the output, and a change from 26.7°C to 15.6°C causes a 14% drop. Cooper (1969a) also concluded that decreasing the average ambient temperature decreases the output. In this case, even though both water and glass temperatures decrease and the difference increases, it is not sufficient to compensate the general fall in the overall temperature of the system. This effect becomes evident at relatively lower temperatures. This is in agreement with the earlier work of Lof *et al.* (1961b). This phenomenon is to a good extent explained by the increase in $h_{c\omega}$ and h_{eff} with increasing $\bar{T}_\omega$.

3.8.3(c). *Effect of Solar Radiation and Loss Coefficient*

The effect of solar radiation (total) and the loss coefficient on the daily output have been shown (United Nations, 1970) in Figs. 3.11 and 3.12 respectively. It is clear from Fig. 3.11 that in determining the output of a still (having different loss coefficients) solar insolation is the single most important parameter. It will depend to some extent upon how the radiation is distributed throughout the day (United Nations, 1970), however, this is a second-order effect, and it is usually sufficient to consider only the total radiation received each day. It is obvious, from the data given in Figs. 3.11 and 3.12, that the best situation for a solar still is one where both the daily insolation (Akinsete *et al.* 1979) and the mean ambient temperature are consistently high. It is also clear from Fig. 3.12 that the loss coefficient has less influence at higher ambient temperatures. The decrease in loss coefficient by a factor of 5 increases the output by about 45%, which is in close agreement with the work of Morse and Read (1968).

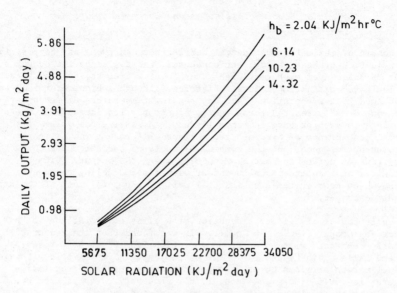

Fig. 3.11. Effect of Solar radiation and loss coefficient on predicted still output at Ambient temperature $T_a = 27°C$ Heat capacity $M_{wo} = 327$ KJ/m²°C Wind velocity $\vartheta = 2.23$ m/sec. (After United Nations, 1970.)

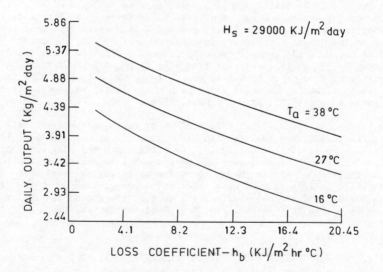

Fig. 3.12. Effect of ambient temperature and loss efficiency on predicted still output at $H_s = 29000$ KJ/m² day. (After United Nations, 1970.)

3.8.3(d). Effect of Double-glass Cover and Cover Inclination (Cooper, 1969a)

The result of the effective thermal barrier between two glass covers, impeding the rejection of heat through the condenser, was to reduce the output by about 25-35%. Even with very high water temperatures the governing factor is the low water-glass temperature difference. Though a high water temperature leads to a high evaporative fraction, the low temperature difference results in a considerably reduced total energy transfer. The increased temperatures also lead to greater base and side losses. Increasing the value of the thermal conductance of air in between glass will result in a slightly improved performance, but the combined effects of radiative and convective heat transfer across the air gap tends to lower the output. From an economical and constructional point of view also, doubling the amount of glass, which constitutes a substantial part of the total cost of the still, is not desirable.

The variation of distillate production with glass cover inclination has been studied by Cooper (1969a) in detail. It was found that the evaporation rate actually decreases with cover slope variation from $0°$ to $45°$, rises at about $60°$ and falls again beyond $75°$. The output changes very little with change of inclination of glass cover; the optimum angle suggested for Indian climates is $10-15°$ (Garg and Mann, 1976). However, a report from U.S.A. states that the angle of inclination has almost no effect on the output (Report No. 190 prepared by Office of Saline Water 1966, USA).

3.8.3(e). Effect of Thermal Capacity on Output

Morse and Read (1968) have studied the effect of thermal capacity of water on the output of the still. They recommended a water depth of about 7.6cm, corresponding to thermal capacity of 1.41 KJ/m^2 for the CSIRO ground-based still (Australia) with bottom insulation and 2.54cm water depth for the still resting on the ground without any insulation. For a daily insolation of 225 KJ/m^2 hr, reducing thermal capacity from 1.41 KJ/m^2 to zero, increases the output by 9%, while increasing from 1.41 KJ/m^2 to 5.64 KJ/m^2 °C (equivalent to 0.3m depth of water) would reduce the output by 7%. It follows therefore that, while the water depth is not a critical parameter, it should be as small as possible (Morse and Read, 1968).

Cooper (1969a) has studied the effect of water depth on the distillate output; the results are indicated in Fig. 3.13. It is obvious that without insulation, the gains from decreasing the water depth are only marginal, but with the insulation the difference is more marked, especially at shallow water depths. A 30% variation was found for water depths from 1.27cm to 30.48cm which is in agreement with the conclusions of Bloemer et al. (1965b). A characteristic of the deep basin still is that the ground temperature varies more smoothly than that in the case of a shallow basin still.

3.8.3(f). Effect of Salt Concentration on Output

The effect of salt concentration on the output of a still has been studied in detail by Baibutaev et al. (1970). Tests were performed on water samples taken from lakes with different salt concentrations. The amount of mineral matter in the lake water was determined by calorimetric and titration methods, using samples taken from different regions of the lake. These samples were then used to prepare water specimens with different salt concentrations ranging from 10% to the saturation value (100% concentration was taken to be the salt concentration in natural water in each lake). The salt concentration in the water was kept constant during the course of operation of the still by adding distilled water and keeping the water level in the drip pan constant (the water layer was held at 10-12mm).

The experiments have shown that as the salt concentration of the water to be distilled increases right up to the saturation point, the output of the still falls linearly and slowly (Fig. 3.14). However, as the salt concentration of the water to be distilled increases, there is an increase in the corrosion damage to the components of the still and, consequently, it then becomes necessary to use materials which are not readily oxidized.

3.8.3(g). Effect of Charcoal Pieces on the Performance of a Still

Akinsete and Duru (1979) have studied the effect of charcoal pieces on performance of a still because of its wettability, large absorption coefficient for solar radiation and their property to scatter, rather than reflect, the solar radiation. It was concluded that the effect of charcoal is most pronounced in the mornings and on cloudy days when the values of direct radiation are low. With high insolation values, the present design is only marginally better. This result implies that the presence of charcoal pieces reduces the start up time and utilizes the diffuse radiation much better than the conventional unlined still. They have also concluded that the charcoal lined still is relatively insensitive to basin-water depth as long as a good amount of the charcoal remains uncovered.

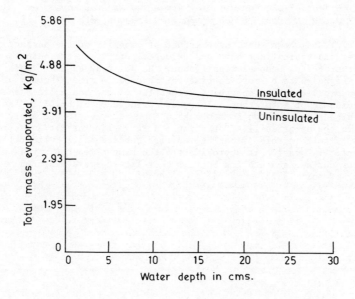

Fig. 3.13. Still production for varying water depths with and without insulation. (After Cooper, 1969a.)

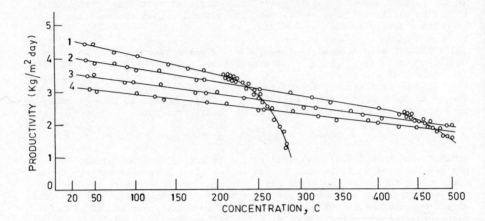

Fig. 3.14. Output of solar still as a function of salt concentration in the initial water (C = 100% for natural water in each lake): 1-4 Lakes Khulis, Tudakul, Makhankul, and Kum-Mazar. (After Baibutaev et al., 1970.)

3.8.3(h). Effect of the Formation of Algae and Mineral Layers on Water and Basin Liner Surface on Productivity of a Still

Cooper (1972) has shown that the presence of deposits on the surface of the basin water and basin liner have a detrimental effect on the output, assuming that no other factor becomes significant. He pointed out that surface reflection appears to be more detrimental than basin liner (bottom of still) reflection because of the absorbing properties of the basin liner except at normal incidence of insolation. There is an increase of 8-15% in the measured albedo as taken on some black surface relative to the scaly surfaces in the solar still. For example a reflection of 21.7% was found by Cooper (1972) with a thin, floating, whitish surface scale. The reflection of an identical location at approximately the same time, but a reasonably clean water surface and a black butyl rubber liner gave a measured reflection of only 13.7%.

4. MULTIPLE EFFECT SOLAR STILLS

4.1. GENERAL CONSIDERATIONS

A multiple effect solar still is a typical high performance still, in that, it produces a larger quantity of distilled water for a given insolation than a single-effect type of the same area. This is accomplished by utilizing the latent heat released by the condensing water vapor, which would otherwise be rejected into the atmosphere, to heat more saline water at a lower temperature. Thus, one essentially has a multi-stage system in which the successive stages yield progressively lesser quantity of distilled water. Naturally, the multi-effect stills involve sophisticated designs, are expensive and operate efficiently only at relatively high temperatures compared to the conventional single-stage still (Howe and Tleimat, 1974; Talbert *et al.*, 1970; Henrik, 1972). The compensation is, of course, the much higher yield of distilled water and if this be taken into account, the capital cost of a standard multi-effect still may become comparable (though higher) to that of a single stage system (Howe and Tleimat, 1974). The multi-effect still would, therefore, be particularly suitable for arid regions, where the land occupied by a distillation unit is at premium on both, economic as well as aesthetic grounds. The ultimate acceptance of multi-effect solar stills will depend on their long term performance and yield during periods of both high and low insolation; yield of distilled water during high solar insolation should compensate the low yield during low insolation. Considerable work on performance of multi-effect stills over extended periods needs to be done before their feasibility is established.

A multi-effect still operates on essentially the same principle as the conventional roof-type still. The transmitted solar energy is absorbed at an absorber. Water in contact with the absorber is heated and consequently, some of it evaporates. The transparent cover of the still is cooler than the absorber or saline water and therefore, the vapors condense on the cover. The latent heat released due to condensation is dissipated by a complex heat transfer mechanism involving conduction, convection and radiation as well.

In what follows some of the proposed designs of multi-effect solar stills will be discussed. In section 4.7 the particular case of double basin still is discussed in detail as regards both theory and experimental performance.

4.2. DIFFUSION STILL (Oltra, 1972)

A schematic representation of a three-effect still, called the Diffusion still, proposed by Oltra (1972), is shown in Fig. 4.1. The system comprises of two separate units. One is a hot water storage tank, coupled to a solar collector, and the other is the distillation unit, which produces the distilled water. The distiller consists of a series of metal plates mounted vertically in a container. Referring to Fig. 4.1, the plate on the extreme left, which we shall call the first plate, is heated by allowing hot water from the storage tank to flow through the tubing bonded to the plate on its left side surface. On the other hand, the plate on the extreme right, which we shall call the last plate, is cooled by allowing saline water (feed water) to flow through the tubing bonded to this plate on its right side surface. The hot water enters the first plate from the top and flows back to the hot water tank from the bottom of the plate. The water flow direction is just reversed in the case of the last plate, but instead of taking it out of the system, the saline water is allowed to gently flow (in a thin layer) down the right side faces of all but the last plate in the distiller.

The operation of the still is as follows. Hot water from the storage tank heats the first plate causing evaporation of the saline water flowing down the right side surface of the plate. The vapors condense on the left side surface of the second (neighbouring) plate, releasing the heat of vaporization into this plate. This causes further evaporation of the saline water flowing down the right side face of the second plate. The process continues until the last plate, in which the condensation of the vapors simply causes a preheating of the in-flowing saline water. The condensed water flows down the plates into collector channels and is taken out of the system. The concentrated brine collects at the bottom of the distiller and can also be removed from the system.

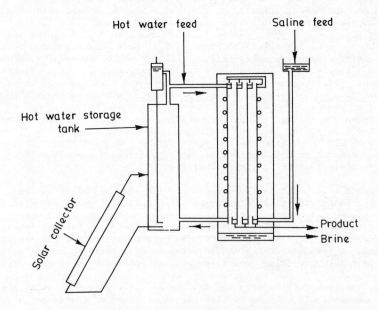

Fig. 4.1. Diffusion solar still. (After Oltra, 1972.)

One can dispense with the hot water storage tank coupled to the solar collector by having a transparent cover for the left side of the distillation unit. In that case, however, the entire system would have to be tilted as in the case of flat plate collectors and several advantages associated with the flow of water down the vertical plates would be lost.

4.3. CHIMNEY TYPE SOLAR STILL (Bartali et al., 1976)

This type of still, developed by Bartali et al. (1976), resembles a conventional still except that a chimney containing a heat exchanger is appended to one end of the still (Fig. 4.2). The saline water enters the chimney, flows through the heat exchanger and then enters the still. Water gets evaporated in the still due to the solar radiation. The vapor diffuses into the chimney and condenses on the external fins of the heat exchanger. The distilled water accumulates at the base of the chimney and can be piped out. The latent heat released in the heat exchanger preheats the in-flowing saline water. The concentrated brine may be removed from the still basin.

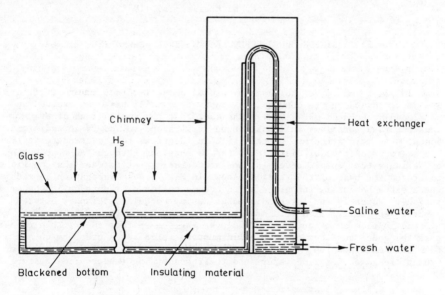

Fig. 4.2. The chimney-type solar still.
(After Bartali et al., 1976.)

4.4. HEATED HEAD SOLAR STILL (Bartali et al., 1976) (Fig. 4.3)

In this design, Bartali et al. (1976) used the concept of having separate units for heat collection and distillation. The distillation unit comprises three metal plates placed parallel to one another but slightly inclined to the horizontal, thereby forming two enclosures. Saline water enters the system through the piping bonded to the upper surface of the central plate. The water is then conducted through the solar collector, where it gets heated, and is then allowed to flow back onto the lower plate of the distiller. Water evaporates here and condenses on the lower surface of the central plate and flows out of the distiller due to gravitational forces. Concentrated

brine accumulates in the lower enclosure of the distiller and can be
conveniently piped out. The heat liberated by the condensing vapors is
utilized to preheat the inflowing saline water.

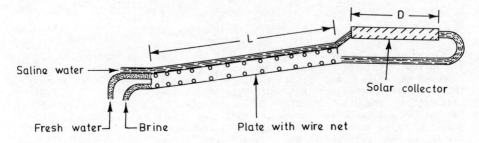

Fig. 4.3. The heated head solar still.
(After Bartali et al., 1976.)

4.5. A SIMPLE MULTI-EFFECT BASIN TYPE SOLAR STILL

4.5.1. Design Principle

With a view to minimize the capital cost of a multi-effect solar still, a
new design based on the conventional basin type still has been evolved and
developed by Lobo et al. (1977). The fact that the sealing and other
constructional problems in a basin type still stand solved is an additional
bonus. The main structure of the still is identical to the standard still.
The extra production of distilled water is obtained through the evaporation
of saline water kept in several transparent basins set between the top glass
cover and the base of the still as shown in Fig. 4.4. Evaporation in each of
these extra basins is energised by the condensation of vapors on its under
surface. Water in the intermediate basins may be held stationary between
several vertically fixed retaining walls in each basin (see Fig. 4.4) or
allowed to flow as a continuous film over the transparent sheets. In either
case the under surfaces of each basin must be wettable in order to obtain a
film like condensation. The basin side walls should be sealed properly to
prevent leakage of saline water onto their condensing (under-side) surfaces.

4.5.2. Construction

Two test models of identical external dimensions were constructed. Expanded
polystyrene was used for the side walls and base of the still. A clear
window glass, of thickness 3mm, formed the top cover, which was inclined at
an angle of $10°$ to the horizontal. The first model was a standard single
basin still. In the second model, a 2mm thick glass sheet was set parallel
to the top cover. This glass sheet had 9 vertical retaining walls separating
10 channels that contained the saline water required to be distilled.
Distilled water yield was measured with a calibrated jar and the temperatures
with mercury thermometers. Incident solar intensity was measured on a
horizontal plane using a Robitsch-Fuess Actinograph. Diffuse radiation was
estimated from the recordings from a shadow band.

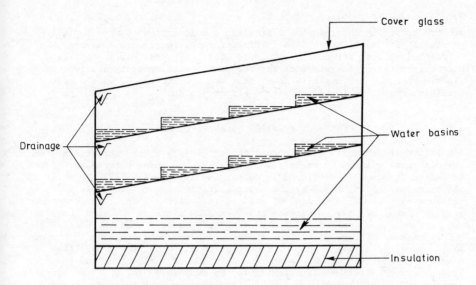

Fig. 4.4. Multiple effect still. (After Lobo et al., 1977.)

The two units were operated simultaneously under identical conditions. Every morning 2.5 Kg of water was poured into the main basin of each module and 0.25 Kg of water into each of the 10 channels of the intermediate basin, corresponding to an average depth of about 1cm, sufficient to avoid occurrence of dry patches even during days of maximum solar radiation. For the double effect test module, separate measurements were made of the condensate collected from the outer glass cover, designated first effect yield, and from the underside of the intermediate basin, designated second effect yield. On some days, hourly measurements of distilled water output were taken along with the mean water temperatures in a conventional (single effect) still and in each effect of the double effect module. Corresponding incident solar energy was obtained by integrating the area under the curves obtained from the actinograph records.

4.5.3. Performance

The main conclusion that can be drawn from this study is that, the gain in the distilled water output of the double effect still over the standard still is at least 40% at 20 MJ/m^2/day and increases with total daily radiation to about 55% at 25 MJ/m^2/day. Since maximum water demand usually occurs on days that receive maximum incident solar energy, the non-linear improvement factor is an advantage, because a plant built to meet maximum demand is less likely to be over-designed for normal conditions.

Lobo's claim that the cost of the intermediate basin would add about 15-25% to the cost of the standard still per unit area, is at best an understatement; however we go along with his projections in this paragraph. At an incident energy of 25 MJ/m^2/day the corresponding yield per unit capital cost would be about 25-35% higher. On the whole, the cost for a given yield would be 20-25% cheaper, an improvement that would justify more research into this type of still, especially since the gain in yield could, perhaps, be made to approach 80% per effect, which is normal in more conventional multi-effect distillation plants (Donald, 1950). Since the operation of the new still is not essentially different from that of the conventional stills, except for the cleaning procedure for the intermediate basins, it should introduce no major adaptation problems. If plastics could be utilized for the intermediate basins, the corresponding cost per unit yield could be further reduced, especially because, by taking advantage of the much smaller thicknesses possible in plastic sheets, substantially lower heat conductance values should prove feasible, resulting in higher second effect yields.

4.6. THREE-EFFECT MULTIPLE SOLAR STILL

A multiple-effect solar still consisting of closely spaced parallel plates (or elements) with vapor-tight sealing so that the space between each forms an "effect", has been developed by Cooper and Appleyard (1967). The first plate (absorber) is heated by the solar radiation and the last plate (condenser) cooled due to flow of cold saline water and thus, a temperature gradient results. Saline water flows down each plate on the side opposite the heat source. Some of the water evaporates, and, due to the partial vapor-pressure difference, diffuses across the space to condense on the cooler plate above. The latent heat of condensation is conducted through the condenser element to promote evaporation in the next effect. Thus the condenser for one effect is essentially the evaporator for the next except for the first and last elements, which are only an evaporator and condenser respectively.

4.6.1. *Construction*

A schematic of the still developed by Cooper and Appleyard (1967) is shown in Fig. 4.5. It comprises of three effects, each consisting of an evaporator and a condenser. The terry towelling wick for each effect was clamped at the discharge end and surrounded a feed-water distributor tube at the inlet end, the position of the tube being adjustable to allow tensioning of the wick if necessary. The clamping of the towel at the discharge end caused the concentrated brine to flow into a catchment channel and then finally, to the brine discharge containers. On top of the condenser for each effect an angle section collected the distilled water. Directly in contact with the wick for each effect was a fine brass wire mesh held in contact by cross threaded nickel-coated copper wire. This provided additional support to the wick. The elements and collector glass were held in a frame which consisted of a laminated wooden base, perspex sides and yellow-pine ends, the whole frame being mounted on an angle-iron box stand that allowed adjustment of the inclination.

The still was fed with saline water from a constant-head tank through a filter-distributor system and three polythene tubes; the flow was controlled by pinch-cocks.

The first three elements were of copper, the final condenser being of galvanized steel for increased strength.

The complete assembled still without insulation measured 1m long and 0.7m wide. The effective collector area was $0.56m^2$. The effective distance between each element was 1.9m. To intercept the maximum insolation over the test period, the still was oriented in a north-south direction at an angle of 45° to the horizontal.

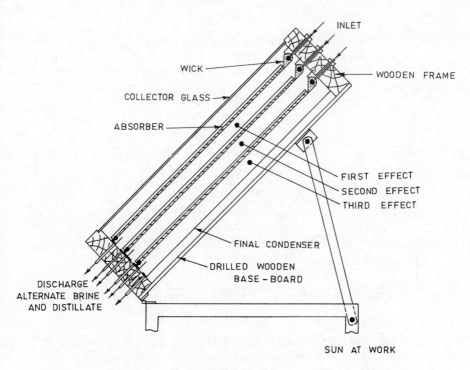

Fig. 4.5. Cross section of still showing constructional detail. (After Cooper and Appleyard, 1967.)

4.6.2. Results

The water yield in Kg per day for the still as a whole is shown in Fig. 4.6. The output varied from 0.3 Kg per day for an insolation of 4100 KJ/m^2 to a maximum of 2.722 Kg per day at an insolation of 130000 KJ/m^2. For a constant feed water flow the distillate output increases with increasing insolation and temperature, thus decreasing the discharge brine flow and hence the sensible heat loss. Due to increased thermal losses at higher insolation and temperatures, it is expected that the increase in distillate output with increasing insolation will tend to flatten out. One of the causes of the scatter of the experimental results was the inability to keep the flow rate and the ratio of flows to each effect constant over long periods. In general, the points above the curve were for flows less than those below the curve.

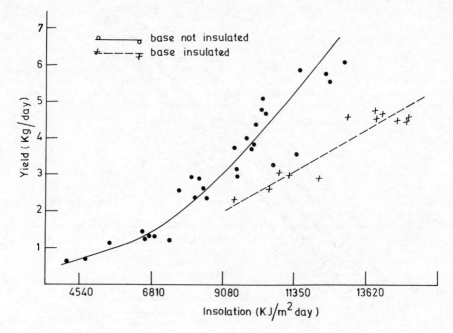

Fig. 4.6. Total water yield per day. (After Cooper and Appleyard, 1967.)

4.6.3. Discussion

The effect of insulating the base of the still was to decrease the output, which is contrary to usual practice with roof-type stills. There was a 34% decrease for 9000 KJ/m² day and 42% for 12000 KJ/m² day. Several reasons were advanced to explain this observation. When the base is insulated, heat loss from the final condenser is minimized, thus raising the temperature at all points within the still. This increased temperature results in greater thermal expansion. It was noticed that for an insolation of 15000 KJ/m² day, vapor was issuing out from a previously undetected leak. If the base is insulated, the temperature gradient through the still should decrease. The partial pressure increases approximately exponentially over the range of temperatures encountered in a multiple effect still, but a point is reached where the temperature difference across an effect decreases to such an extent that the partial-pressure difference decreases, lowering the still output.

With increased still temperature, losses to the atmosphere through the sides also increase. The emissivity of the selective surface, if any, also increases, reducing the effective collection of incident energy. This is supported by Fig. 4.6 where as the insolation decreases, so does the difference between the yield curves for the base uncovered and covered.

4.7. DOUBLE BASIN SOLAR STILL

4.7.1. *Background*

Lobo and Araujo (1977) have proposed a multiple-effect double basin solar still. However, they have not discussed the theoretical aspects and the design considerations in any detail. Moreover, the still proposed by them is difficult to fabricate, operate and maintain. This section discusses a simpler design, proposed by Malik *et al.* (1978), for the double basin still; Fig. 4.7 shows a schematic of this still. The novelty in this design is that the latent heat released at the lower surface of the inner glass cover by condensation of vapor from the lower basin (the still has two glass covers) is utilized to heat a thin layer of water on the upper surface of the same cover. This again causes evaporation of water and the vapors from the upper basin condense on the second (outer) glass cover. On a typical winter day in Delhi the distillate output from such a still was observed to be about 36% higher than that from a still with a single glass cover of the same area. Following Sodha *et al.* (1980c) the performance of the double-basin still may be analysed using the periodic theory developed earlier. This analysis along with comparison of the experimental measurements and theory has been presented in section 4.7.3.

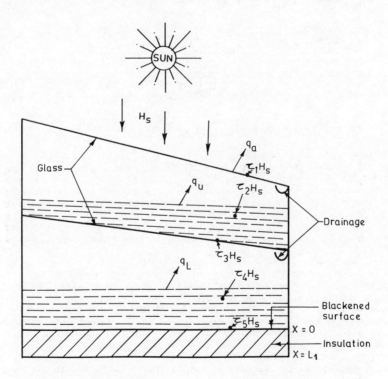

Fig. 4.7. Double basin still. (After Sodha *et al.*, 1980c.)

4.7.2. Construction

The double basin still proposed by Malik et al. (1978) differs from the conventional still in having another transparent sheet (glass is preferred for well known reasons) fixed in between the basin liner and glass cover; this sheet serves as the base of an extra basin for the saline water. The water in the upper basin reduces solar intensity reaching the base of the still, but makes use of the upward heat loss by the water in the lower basin. Thus, in effect, the whole assembly behaves as two single basin stills kept one above the other and as a result, it has a smaller area requirement per unit mass distilled as compared to a single basin-still.

4.7.3. Analysis*

The major heat flux components are shown by arrows in Fig. 4.7. Energy balance conditions for the top glass cover, middle basin water, middle glass cover, lower basin water and absorbing surface can, respectively, be written as

$$M_{gu} \frac{dT_{gu}}{dt} = \tau_1 H_s + q_u - q_a \tag{4.7.1}$$

$$M_{\omega u} \frac{dT_{\omega u}}{dt} = \tau_2 H_s + h_5(T_{gL} - T_{\omega u}) - q_u \tag{4.7.2}$$

$$M_{gL} \frac{dT_{gL}}{dt} = \tau_3 H_s + q_L - h_5(T_{gL} - T_{\omega u}) \tag{4.7.3}$$

$$M_{\omega L} \frac{dT_{\omega L}}{dt} = \tau_4 H_s + h_3(\theta|_{x=o} - T_{\omega L}) - q_L \tag{4.7.4}$$

$$\tau_5 H_s = h_3(\theta|_{x=o} - T_{\omega L}) - K_1 \frac{\partial \theta}{\partial x}\bigg|_{x=o} \tag{4.7.5}$$

where

$$q_a = h_2(T_{gu} - T_a)$$
$$q_u = q_{r\omega u} + q_{c\omega u} + q_{e\omega u}$$
$$q_L = q_{r\omega L} + q_{c\omega L} + q_{e\omega L}$$

u and L refer to the upper and lower glass covers, respectively, and $q_{c\omega u}$, $q_{e\omega u}$ and $q_{r\omega u}$, $q_{c\omega L}$, $q_{e\omega L}$ and $q_{r\omega L}$ are same as $q_{c\omega}$, $q_{e\omega}$ and $q_{r\omega}$ of equations 2.1.7, 2.1.19 and 2.1.20 provided T_ω and T_g are replaced by $T_{\omega u}$ and T_{gu}, and $T_{\omega L}$ and T_{gL} respectively; $\theta(x,t)$ is the temperature distribution below the absorbing surface and is governed by the heat conduction equation i.e. 3.4.15.

The energy balance at the surface of insulation in contact with air is given by (the effect of box-sheet thickness is neglected)

$$-K_1 \frac{\partial \theta}{\partial x}\bigg|_{x=L} = h_4(\theta|_{x=L} - T_a) \tag{4.7.6}$$

---

*After Sodha et al. (1980c).

Multiple Effect Solar Still

Assumptions similar to those made in section 3.4 have been made in writing Eqns. (4.7.1)-(4.7.5). Using the expression for P given by Eqn. (3.4.4), one obtains

$$q_u = h_1(T_{\omega u} - T_{gu}) \tag{4.7.7(a)}$$

and

$$q_L = h_6(T_{\omega L} - T_{gL}) \tag{4.7.7(b)}$$

where h_1 and h_6 are evidently temperature dependent quantities; but since the range of variation of the temperature of water and glass is quite small h_1 and h_6 may be considered to be constant and can be evaluated from Dunkle's (1961) relation (section 3.4.1(a)), with temperatures T_ω and T_g replaced by their mean values. Similarly, one can write

$$q_a = h_2(T_{gu} - T_a) \tag{4.7.8}$$

where, h_2 is considered to be a constant and is calculated making use of the mean of T_{gu} and T_a as discussed in section 3.4.1(a).

Equations (4.7.1)-(4.7.4) can now be rewritten as

$$M_{gu}\frac{dT_{gu}}{dt} = \tau_1 H_s + h_1(T_{\omega u} - T_{gu}) - h_2(T_{gu} - T_a) \tag{4.7.9}$$

$$M_{\omega u}\frac{dT_{\omega u}}{dt} = \tau_2 H_s + h_5(T_{gL} - T_{\omega u}) - h_1(T_{\omega u} - T_{gu}) \tag{4.7.10}$$

$$M_{gL}\frac{dT_{gL}}{dt} = \tau_3 H_s + h_6(T_{\omega L} - T_{gL}) - h_5(T_{gL} - T_{\omega u}) \tag{4.7.11}$$

$$M_{\omega L}\frac{dT_{\omega L}}{dt} = \tau_4 H_s + h_3(\theta|_{x=o} - T_{\omega L}) - h_6(T_{\omega L} - T_{gL}) \tag{4.7.12}$$

The solar intensity and ambient air temperature can be considered to be periodic and are then given by Eqn. (3.4.17)..

In view of Eqn. (3.4.15), one can assume the following periodic solution:

$$\theta(x,t) = Ax + B + \text{Re} \sum_{n=1}^{\infty} \{C_n \exp(-\beta_n x)$$

$$+ D_n \exp(\beta_n x)\} \exp(in\omega t) \tag{4.7.13(a)}$$

$$T_{gn} = g_o + \text{Re} \sum_{n=1}^{\infty} g_n \exp(in\omega t) \tag{4.7.13(b)}$$

$$T_{gL} = G_o + \text{Re} \sum_{n=1}^{\infty} G_n \exp(in\omega t) \tag{4.7.13(c)}$$

$$T_{\omega u} = h_o + \text{Re} \sum_{n=1}^{\infty} h_n \exp(in\omega t) \tag{4.7.13(d)}$$

$$T_{\omega L} = H_o + \text{Re} \sum_{n=1}^{\infty} H_n \exp(in\omega t) \tag{4.7.13(e)}$$

where

$$\beta_n = n^{1/2} \alpha (1 + i)$$

and

$$\alpha = \left(\frac{\omega \rho c}{2K}\right)^{1/2}$$

The constants A, B, C_n, D_n, g_o, g_n, h_o, h_n, G_o, G_n, H_o and H_n can be determined by substituting Eqn. (4.7.13) in Eqns (4.7.9-4.7.12), (4.7.5) and (4.7.6) and solving the resulting system of equations.

The heat flux corresponding to evaporation from the upper basin is given by

$$q_{e\omega u} = h_{eu} (T_{\omega u} - T_{gu})$$

while, from the lower basin, it is

$$q_{e\omega L} = h_{eL} (T_{\omega L} - T_{gL})$$

The amount of water distillate per unit time per unit basin area is given by

$$\dot{m}_e = (q_{e\omega u} + q_{e\omega L})/\mathcal{L} \qquad (4.7.16)$$

In the temperature range of interest $\mathcal{L}$ may be treated as a constant.

4.7.4. Comparison with Experiment

An experimental still (Sodha et al., 1980c), having a basin area of 0.9m × 0.8m, was fabricated at IIT Delhi out of galvanized iron sheet (24 SWG) and was encased in a wooden box; the thickness of insulation (glass wool) between the bottom and walls of the basin and the box was 0.05m. The vertical heights of the outer glass cover of the still were 0.29m and 0.15m with a slope of $10°$ along the breadth of the still. The basin surface was painted black using black-board paint. A 3mm thick glass cover was fixed on the top by means of a frame made of aluminium 'T'. Another glass sheet was fixed similarly, with a slope of $7°$ at a height of 0.10m from the lower basin along the shorter side of the still. Care was taken to make the system air-tight with the help of rubber gaskets. V-drains of galvanised iron sheet were used for the drainage of distilled water in the two compartments. A slight slope was given to the drainage systems so as to enable the distillate to flow out easily. For a comparative study of the performance of this still, a single basin still of identical outer dimensions was also fabricated.

Equal amount of water (to a depth of 6cm) was poured into the lower basin of double basin still and into the single basin still, while in the upper basin of the former still, water was filled in so as to just cover the glass sheet. The stills were exposed to the sun for 3 days so as to simulate the periodic conditions. Amount of distilled water, solar intensity and ambient air temperature were recorded at hourly intervals on 16 March 1979 at IIT Delhi. Temperatures of the glass cover and basin water were measured by using a calibrated copper-constantan thermocouple with the help of a D.C. microvoltmeter; other junctions of the thermocouples were kept in an ice bath. The measured distilled water yield was found to be in reasonable agreement with the theoretical estimates.

The main conclusions of the above study are:

1. The daily distillate production of a double-basin still is, typically, 36% higher than that of a single basin still.

2. An inspection of the expression (4.7.16) for daily distillate yield, shows that the daily distillate production is independent of the water mass in the basin.

3. It is observed that the productivity increases rapidly with increasing insulation thickness up to 4cm and it then increases rather slowly.

4. The periodic analysis of the double basin still is in reasonable agreement with experiments. The uncertainty in the values of the chosen parameters may also be partially responsible for the absence of a closer agreement between theory and experiment.

Transient performance of double basin solar still has also been studied by Sodha *et al.* (1980e).

5. INCLINED SOLAR STILLS

A single basin solar still suffers from some drawbacks; the horizontal surface of water (except at locations near the equator) intercepts lesser solar radiation than a surface which is tilted appropriately.* The productivity of a conventional still is also limited by the large thermal capacity of the water in the basin. Thus the efficiency of solar stills can be increased by

1. having the liquid surface oriented at an optimal inclination to receive maximum solar radiation,

2. placing the transparent glass cover of the still parallel to the water surface (this minimizes reflection losses) and

3. exposing lesser amount of saline water at a time to solar radiation to get high temperature of water (T_w) (a smaller mass of water heated to higher temperatures evaporates more rapidly).

The main difficulty encountered in designing tilted solar stills is the problem of maintaining an inclined water surface. Several ingenious methods have been tried and will be discussed in the following sections of this chapter. It is worthwhile to mention here that improvements carried out along the lines indicated in this chapter can increase the efficiency of a still by as much as 75%.

5.1. TILTED TRAY OR INCLINED-STEPPED SOLAR STILL

A well known example of a tilted solar still is the inclined-tray solar still, constructed and studied by Tleimat and Howe (1966) at the University of California (Fig. 5.1). The distilled water yield (as a function of time) is shown in Fig. 5.2 for the tilted still as well as for a conventional basin

---

*For maximum integrated solar insolation in a year the surface should face south and be inclined to the horizontal at an angle equal to the latitude of the place.

type still of same area. It is seen that the yield of the tilted still is much larger than that of the conventional still during sunshine, but the nocturnal production of the tilted still is almost zero. This difference in performance is partly due to the difference in the water depths and hence in the heat capacities in the case of the inclined tray and the deep basin still. The deep basin still, with water to a depth of 0.3m, had a very large heat capacity and consequently, the solar energy increases the water temperature to a maximum of only about 40°C. Distillation at this low temperature proceeds slowly, resulting in a correspondingly low decrease in water temperature due to its large heat capacity; thus the nocturnal production is significant and accounts for a fairly large fraction of the total. In contrast, the tilted-tray unit, with less than 1.27cm depth of water, shows maximum water temperatures greater than 66°C and has a correspondingly high rate of distillation during sunny hours. These high rates cause a rapid cooling of the water in the basin, so that soon after sunset, the temperature of the water becomes too low to cause evaporation and nocturnal distillation is very little.

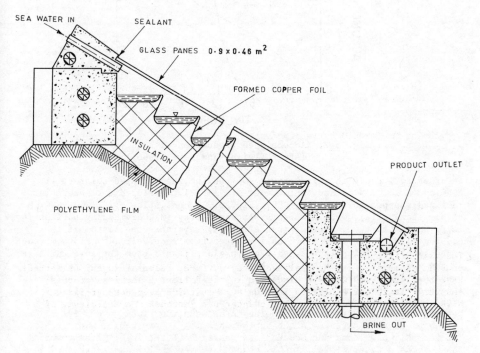

Fig. 5.1. Cross section of tilted tray solar still.
(After Tleimat and Howe, 1966.)

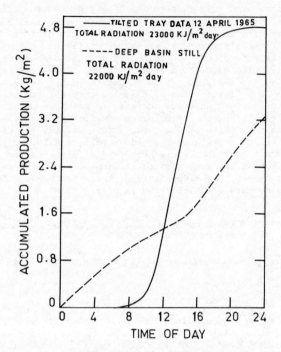

Fig. 5.2. Accumulated production of solar stills.
(After Tleimat and Howe, 1966.)

In order to improve the productivity of the inclined-stepped solar still, Achilov *et al.* (1976) and Akhtamov *et al.* (1978), developed a regenerative inclined-stepped solar still (RSS) (Fig. 5.3), consisting of a housing with double glazing (2), within which a blackened tray (3), divided by baffles (4), is fixed. A layer of thermal insulation (5) separates the bottom of the housing (1) and the tray (3). Saline water from reservoir (10) flows through pipes into the enclosure between the two glass covers from the lower edge (6) of the still; the entire still is kept at a lower height than the reservoir. After filling up this space water flows out into the basin (3) of the still. The channels formed by the baffles hold limited quantities of water, but they enable the tilting of the still as a whole at an angle of $30°$ to the horizontal. The saline water flowing in between the glass covers is preheated to some extent. After further heating in the basin, water vapors are formed, which condense on the lower surface of the inner glass cover. The condensed water flows into the condensate trough (9) and is then piped out of the still.

An experimental model of the RSS was fabricated and tested at the Bukhara State Pedagogical Institute between 1st June and 31st August 1978. The overall distillation was observed to be 20 litres/day. It was seen that the still was more efficient during the afternoons. The productivity was also found to be about 1.5 times that of the single slope inclined stepped still.

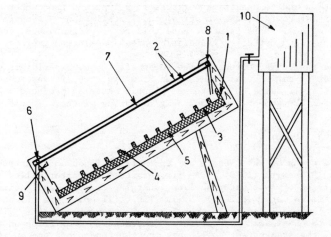

Fig. 5.3. RSS System. (After Akhtamov *et al.*, 1978.)

5.2. STEPPED INCLINED STILL (Achilov *et al.*, 1973b)

Three stills of the step-inclined type were constructed by Achilov *et al.* (1973b) with a total evaporation area of 1.6m^2 each, using the FK-20 VTU MKhP 652-55 foamed plastic and the P2020-T polyethylene. The stills were tested at the solar testing area of the Bukhara State Education Institute and during expeditions to the Kyzyl-Kum Desert undertaken by Gazlineftegaz-Bukharneftegaz.

The basic components of the still are a foamed plastic pan separated into several sections by steps perpendicular to the base of the still, a polyethylene bath, a cover transparent to insolation (window glass), and a rigid base. The system is shown schematically in Fig. 5.4.

Commercially available foamed-plastic slabs having an evaporation area of 0.4m^2, were fixed on a single rigid base. The bath was 4mm thick and placed inside the foamed plastic body, separated from it by an air gap. The material had sufficient mechanical strength in the temperature range of interest.

The experiments showed that, at temperatures of 60-75°C and moisture content around zero, the foamed plastic sheet does not change its mechanical properties. However, the moisture content of the medium played an important role.

The main parameters of this particular still design are the optimum values of the step height, the number of steps and the distance between the evaporation and condensation surfaces. To determine the optimum distance between the evaporation and condensation surfaces, Achilov *et al.* (1973b) constructed stills in which this separation was 6.9 and 12cm and to determine the number of steps and their height they investigated stills with 8, 12 and 16 steps per meter width with heights of 2, 4 and 6cm.

Extensive experiments and calculations showed that when the distance between the evaporation and condensation surfaces is 9-10cm and the number of steps is 10-12 per meter width, the output of the installation is a maximum at a step height of 4cm. This is so because this particular chamber geometry ensures the maximum effective evaporation area and minimum inertia.

To determine the annual output of the solar still, the experiments were extended over a year and, as can be seen from Fig. 5.5, it was found that the portable still based on the foamed plastic material produced 5-6 Kg/m^2 per day during the summer and 1.5-2.3 Kg/m^2 during winter.

5.3. MULTI BASIN SYSTEM

Two separate types of multi-basin systems were designed and constructed by Moustafa and Brusewitz (1979). The first system, hereafter called system No. 1, consisted of salt water shelves, charging pump and collection tray for the distilled water (see Fig. 5.6). The still cover was oriented at a 45° angle. The projected basin area for the system was $0.836m^2$. The depth of water in the basin was kept at 5cm. Condensation took place on the underside of the transparent collector, and the condensate accumulated in the product tray. The second system, hereafter called system No. 2, had similar stepped shelves and a charging pump plus a condenser reservoir on the shaded side of the still which extended above the top most saline water shelf (see Fig. 5.7). The projected basin area for this system was also $0.836m^2$, the cover was oriented at 45° angle, and the depth of water in the basin was maintained at 5cm. Condensation took place on both the underside of the cover and that of the condenser reservoir. In both basin type systems the temperatures of the liquid in the evaporation trays, air vapor mixture, transparent cover and condensate were recorded. Solar insolation and ambient air temperature were also measured.

5.3.1. Basin-type Solar Still

It was noticed that, system No. 2 operated at lower temperature due to the presence of the condensation reservoir on the shaded side of the system. The incident solar radiation and hourly productivity for the two systems were also recorded. In both the systems, the productivity lagged the incident radiation curve except for the morning period where the productivity rises and then drops as the ambient air temperature begins to rise shortly after sunrise.

In system No. 2 the presence of the condenser reservoir did not appear to have any dramatic effect on the system performance, with a relatively small percentage of the condensation taking place on the underside of the reservoir and the majority of the condensation on the underside of the transparent cover.

Calculations of the energy utilization by the systems were conducted by Moustafa *et al.* (1979) and it was found that the calculated efficiency is comparable to the observed efficiency.

Inclined Solar Still

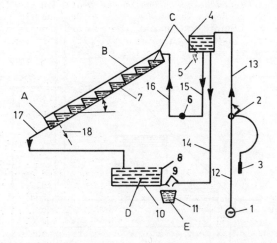

Fig. 5.4. Schematic illustration of the OSP-100 solar still: (A) distillate; (B) glass; (C) initial water; (D) distilled water; (E) drinking water; (1) pump; (2) 3-way tap; (3) nozzle; (4) tank for water to be distilled; (5) draining tap; (6) cut off valve; (7) still; (8) distilled water tank; (9) mixing tap; (10) tap; (11) measuring bucket piping; (12,13) rising; (14) draining for initial water; (15) links for items 4 and 6; (16) links for items 6 and 7; (17) draining for distillate; (18) waste.

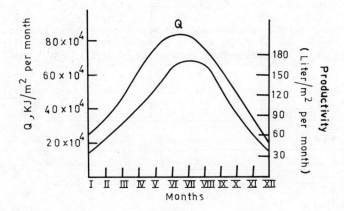

Fig. 5.5. Monthly sums of solar radiation and output of the portable solar still. (After Achilov *et al.*, 1973b.)

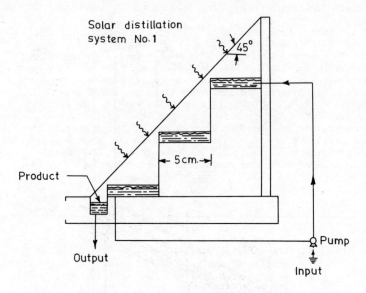

Fig. 5.6. Basin type stepped solar still equipped with charging system. (After Moustafa and Brusewitz, 1979.)

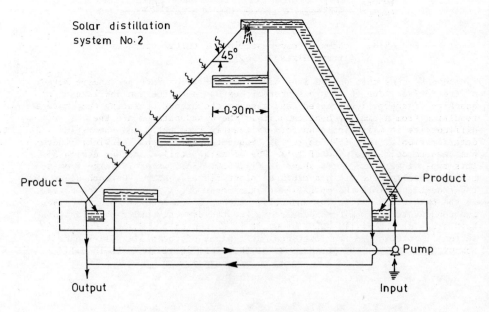

Fig. 5.7. Basin type stepped solar still equipped with charging system and condenser reservoir on the shaded side of the still. (After Moustafa and Brusewitz, 1979.)

5.4. TILTED SINGLE WICK STILLS (After Lof, 1966)

Tilting of the water surface can be simulated by providing a porous water absorbing pad which is mounted in a frame that can be tilted to face the sun more favourably (Telkes, 1956). A diagram of a typical tilted wick distiller is shown in Fig. 5.8. Most of the experimental stills of this type are comparatively small, maximum areas being about $2.3m^2$. Salt water is allowed to flow slowly from a distributor along the upper edge of the porous wick, usually constructed of some type of black fabric. Solar energy is absorbed in the cloth and evaporation takes place. Condensate over the transparent glass or plastic cover gets collected in a trough at the lower edge of the cover. Unevaporated brine drips from the lower edge of the wick. Beneath the wick there is a suitable waterproof structural material and insulation; side enclosures and supports are also provided.

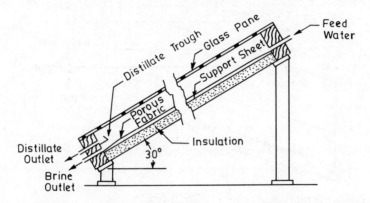

Fig. 5.8. Schematic cross section of tilted wick still. (After Telkes, 1956.)

The advantage of this design is a higher distillation rate per square meter of the surface, due to a more favourable exposure to the sun and larger quantity of energy intercepted, and also to the higher operating temperature resulting from a smaller heat capacity. The disadvantages are the difficulties in maintaining uniform salt-water feed rates and the rapid deterioration of the wick due to the frequent drying up of the wick. There has been no commercial use of this type of solar still. Another design of this type of solar still is shown in Fig. 5.9. Salt water is taken from a trough by collecting it in a cloth of plant fibres (jute). On both sides of the trough the cloth runs parallel to the cover until it reaches the bottom of the salt water collecting channels, which are situated at both the sides. Two condensate collection channels are located directly under the glass at both sides of the still. Obviously, it is economically advantageous to put thermal insulation in the lower and outer parts of the still (Frick and Sommerfeld, 1973). The performance of the still is given in Table 5.1.

TABLE 5.1

Season	Output Kg/m² day	Efficiency (%)	Remarks	Place
December 1963	4.4	46	without insulation	Chile
December 1964	3.8	40	with insulation (when the cloth was very discoloured)	Chile

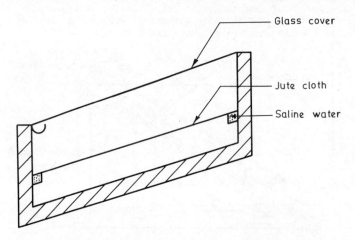

Fig. 5.9. Porous wick type inclined solar still. (After Frick and Sommerfeld, 1973.)

5.5. MULTIPLE-LEDGE TILTED STILLS (After Lof, 1966)

Another variety of small solar distiller, having limited commercial use, combines the efficiency of the tilted model with the simplicity of the basin type (MacLeod and McCracken, 1961). In an improved form of this still, shown in Fig. 5.10, a tilted, shallow, glass-covered box contains a stepped series of shallow, narrow horizontal trays. Salt water is fed to the upper tray, from which it overflows to the next, and so on to the lowest tray and out to the drainage pipe for brine. Solar energy absorbed in the water and on the black bottoms of the trays supplies the heat for evaporation, and moisture condenses on the cover. The condensate is collected in a trough at the lower edge of the cover. Corrosion is minimized in the latest design of this distiller by the use of enameled metal trays (McCracken, 1965). The principal use of this unit has been in providing potable water for individual households having access to brackish sources. Although productivity is somewhat lower than that with the wick type, it appears that there are fewer operating and maintenance difficulties.

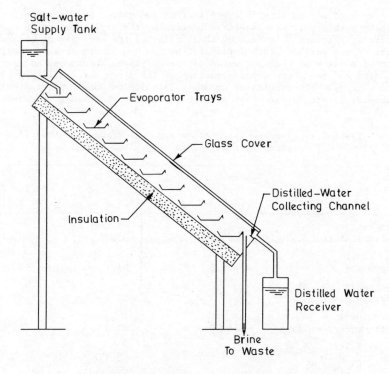

Fig. 5.10. Schematic cross section of multiple-ledge tilted still. (After MacLeod and McCracken, 1961.)

5.6. WICK TYPE COLLECTOR-EVAPORATOR STILL

Apart from the two basin type tilted stills discussed earlier in section 5.3, Moustafa and Brusewitz (1979) also designed and fabricated a wick type solar still (see Fig. 5.11). This system, hereafter called system No. 3, was provided with a water trickling system controlled by a flow regulator and shut-off valve operated by a photocell. The collector-evaporator synthetic wettable mat, 2.54cm thick, was lined with black plastic on the underside. The projected area of the collector-evaporator was $0.182m^2$. The trickle system provided three to four times the productivity of the still to avoid salt accumulation on the collector-evaporator. The still cover was oriented at the same angle as in the system No. 1. Comparative study of these three types of still has been tabulated in Table 5.2 from which it is clear that wick type collector-evaporator performs better than the basin type still.

5.7. SIMPLE MULTIPLE WICK SOLAR STILL

In this section, following Sodha *et al.* (1981a), the design, analysis and experimental performance of a multiple wick tilted solar still is presented. Instead of a thin layer of flowing water, wet jute cloth pieces, dipped at the upper edge in a saline water tank, are placed on the base of a still tilted at such an angle as to intercept maximum solar radiation. The jute cloth pieces of uneven lengths are blackened and placed one upon another, separated by thin polythene sheets. Suction by the capillary action of the cloth fibre, provides a surface of the liquid and the arrangement ensures that the entire surface cloth irradiated by the sun is wet at all times; the portion of a piece of cloth, covered by the polythene sheet does not suffer evaporation and hence the exposed portion of the piece (of small length) retains wetness. The results of an analysis based on Dunkle's relation (cf. Chapter 2) are in excellent agreement with the observed performance of the still.

TABLE 5.2. Average Temperature and Performance for the Three Systems - 20th July, 1977

	System No. 1 (basin) Fig. 5.6	System No. 2 (basin and rev) Fig. 5.7	System No. 3 (wick type)
Average ambient air temperature, $^\circ$C	33.10	33.1	33.1
Average cover temperature, $^\circ$C	36.6	37.1	40.6
Average basin temperature, $^\circ$C	43.3	50.3	44.9
Total incident radiation, KJ/cm^2 day	26700	26660	26660
Total distillate Kg/m^2 day	2.701	2.557	0.127
System overall efficiency, %	24.51	23.20	58.3

5.7.1. *Experiment*

Fig. (5.12b) schematically represents the cross-sectional view at xx of the still shown in plan (Fig. 5.12c). A thin sheet of foam (of thickness 2.5mm) stretched over a net of nylon ribbon 'S' cover a frame made up of aluminium angles and measuring 0.86m x 1.00m forms the base of the still. A window glass cover of 3mm thickness (Gs) is placed above 3cm high walls of insulated foam F (of thickness = 3.5cm). The inner surfaces of the base and the walls of the still are covered with a thin black polythene sheet (P), shown by thin dotted line in Fig. (5.12b), to make it waterproof. The use of rubber gaskets (G), shown by thick dotted line, ensures a good contact of glass with the walls, whereas a frame made of aluminium 'T' tightens the glass to make the assembly airtight. A V-drain of aluminium (O_d) is used for the drainage of the distillate and is slightly sloped so as to enable the distillate to flow out due to gravity. A water reservoir in the form of a small channel, made out of galvanized iron sheet (R), is fixed to the one end of the aluminium base frame. The water surface is realized by means of the wet blackened jute cloth pieces (J) (separated by black polythene sheet) of increasing length spread over the waterproof base with their top edges immersed in the water (W) contained in a reservoir (R). The system is oriented towards south and is kept at an inclination of 15° to receive maximum solar radiation.

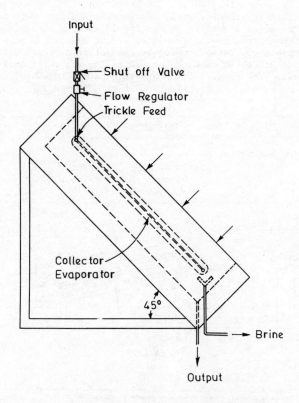

Fig. 5.11. Side view of the wick type collector evaporator system. (After Moustafa and Brusewitz, 1979.)

The capillary action of the jute fibre sucks water upwards from the reservoir and after passing the maximum height, the water rolls down the cloth length under gravity. However, the restricted supply of water by capillary action feeds only a limited length of the cloth, depending upon the rate of evaporation from the wet surface under given conditions. This optimum length of the cloth wetted depends mainly on the solar insolation. Therefore, several cloth pieces of successively increasing lengths are arranged in layers as shown in Fig. 5.12b; these are used to cover the entire base of the solar still. These cloth lengths are separated from one another by thin polythene sheet (P) so that each cloth piece can independently feed its own surface area exposed to the sun. Constant water level is maintained in the reservoir by a constant flow arrangement through inlet I. No thermal insulation (except foam base and walls) was used in the construction of the still. A drainage (O) made of copper pipe, of thickness 0.3mm, is fixed below the drainage of the distillate to enable the extra saline water to come out from the cloth when solar radiation is small in magnitude. The set up is pictorially illustrated in Fig. 5.13.

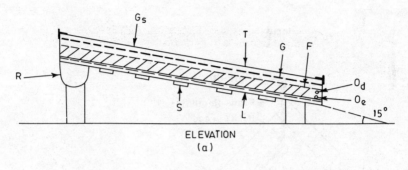

ELEVATION
(a)

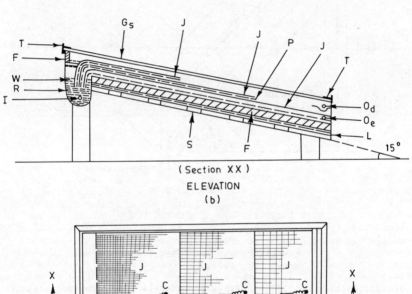

(Section XX)
ELEVATION
(b)

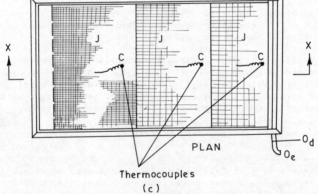

PLAN

Thermocouples
(c)

Fig. 5.12. Schematic representation of experimental configuration. (After Sodha *et al.*, 1981a.)

Fig. 5.13. Simple multiple wick solar still: pictorial view.

In order to measure the temperatures of wetted jute cloth pieces and the glass cover, one end of the thermocouples is attached to every jute cloth piece and at the top of the glass cover, and the other junction is kept in an ice bath; the attachment to the jute cloth is made by placing the end in between the cloth piece and a tiny patch sewn to it. The water surface temperature was taken as the arithmetic average of the readings sensed by several thermocouples, taken with the help of a potentiometer at hourly intervals. The amount of distillate and solar intensity were also recorded at hourly intervals. The hourly variation of solar intensity and ambient temperature on 6th February, 1980, at Delhi, India have been shown in Fig. 5.14. The distillate collected over an interval (one hour) should be related to the mean solar insolation, atmospheric temperature, water temperature and glass cover temperature over the same interval.

5.7.2. Analysis

Using Dunkle's relations (viz. expressions for $q_{c\omega}$, $q_{e\omega}$ and $q_{r\omega}$ given in Chapter 2) and neglecting the heat capacity of the irradiated mass of water and the glass cover, the energy balance for the glass cover and the water sheet may be expressed as

$$\tau_1 H_s + q_{r\omega} + q_{c\omega} + q_{e\omega} - q_a = 0 \qquad (5.7.1)$$

and

$$\tau_2 H_s - q_{r\omega} - q_{c\omega} - q_{e\omega} - h_b (T_\omega - T_a) = 0 \qquad (5.7.2)$$

where $q_{c\omega}$, $q_{e\omega}$, $q_{r\omega}$ and q_a are same as Eqns (2.1.7), (2.1.19), (2.1.20) and (2.2.3) respectively, and

$$\frac{1}{h_b} = \frac{L}{K} + \frac{1}{h_1}$$

The amount of water distillate per unit time per unit area is given by Dunkle (1961)

$$\dot{m}_e = \frac{q_{e\omega}}{\mathcal{L}} = 16.273 \times 10^{-3} \left[\frac{q_{c\omega}(P_\omega - P_g)}{\mathcal{L}(T_\omega - T_g)} \right] \qquad (5.7.3)$$

where T_ω and T_g can be evaluated from simultaneous solution of Eqns (5.7.1) and (5.7.2) by numerical methods; the corresponding saturation vapour pressures can be obtained from the steam table (Steam Table prepared by Ernst Schmidt 1969).

The efficiency of the still is given by

$$\eta = \frac{\int q_{e\omega}\, dt}{\int H_s\, dt} \qquad (5.7.4)$$

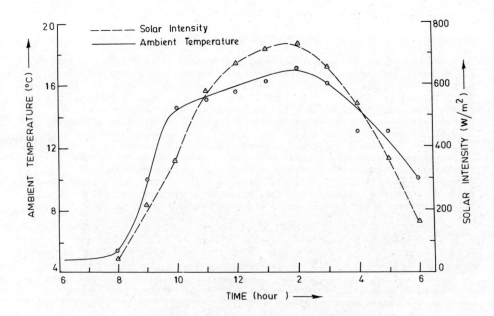

Fig. 5.14. Hourly variation of solar intensity and ambient temperature on February 6, 1980 at Delhi, India. (After Sodha *et al.*, 1981a.)

5.7.3. *Numerical Results and Discussion*

Since the analysis is based on Dunkle's (1961) relation, valid so far for free water surfaces only, it was considered necessary to verify Dunkle's relations for the case of wet cloth surfaces; this can be readily done from the hourly observations of distillate output and temperatures of water surface and glass cover. Figure 5.15 (solid line) illustrates the variation of the rate of distillate output with time, as calculated from Dunkle's relation, using the observed temperatures of the water surface and glass cover by means of the continuous curve; the observed rates of distillation are shown by circles. The excellent agreement between the solid line curve and the circled points validates the use of Dunkle's relations for the present configuration; the other "theoretical curve" is discussed later.

The following parameters have been used in evaluating the water surface and glass temperature from the observed values of solar radiation and atmospheric temperature by the numerical solution of Eqns (5.7.1) and (5.7.2):

τ_1 = 0.05

τ_2 = 0.85 (best estimate from available measurements)

L = 0.0025m

K = .024 W/m °C (ASHRAE, 1967)

h_i = 22.88 W/m² °C (ASHRAE, 1967)

h_{ca} = 14.14 W/m² °C corresponding to a wind speed of 8 Km/hr (Duffie, 1974)

$\mathcal{L}$ = 2372.52 KJ/Kg (Data Tables)

The calculated variation of water and glass temperature, is shown in Fig. 5.16 by continuous and broken curves respectively, and the corresponding observed values have also been shown by △ and ⊙ respectively. It is seen that the theoretical calculations are in good agreement with the experimental results.

The time variation of the water distillate per unit area has been evaluated with the help of Eq. (5.7.3) using the calculated water and glass temperature and is shown in Fig. 5.15 by broken ("theoretical") curve; the results are in reasonable agreement with experiments.

On a typical cold sunny day in Delhi (viz. February 6, 1980) the distillate output was 2.5 Kg/meter² day, corresponding to an efficiency of 34%. The still costs less than half the cost of a basin type still of the same area. Also, the overall efficiency of the basin type still is 30% or lower (Moustafa *et al.* (1979)).

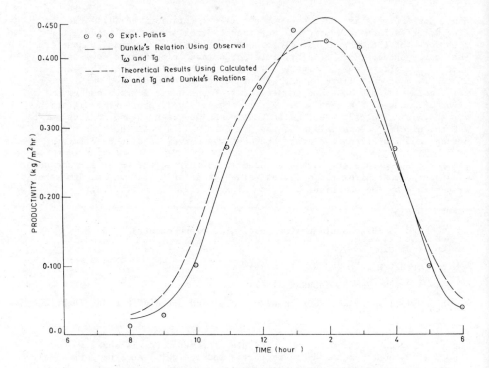

Fig. 5.15. Hourly variation of distillate yield on February 6, 1980 at Delhi, India. (After Sodha et al., 1981a.)

5.7.4. Advantages of the Multiple Wick Still

Some of the noteworthy advantages in a multiple wick still are listed below:-

1. It is light and hence easily portable. A collapsible model which can be neatly folded, can also be developed.

2. It costs less than half of the cost of a basin type still of the same area.

3. The distillate output is significant even on a cloudy day; on a fully cloudy day in winter in Delhi the output was little more than 1 Kg/m^2 day.

4. The water surface on the cloth can be oriented at any optimum angle to receive maximum solar insolation.

5. There is no shadowing effect due to the small height of the side walls.

6. The silt formation occurring on blackened cloth can be brushed off easily or a black dye injected in saline water can alleviate the effect of the silt formation.

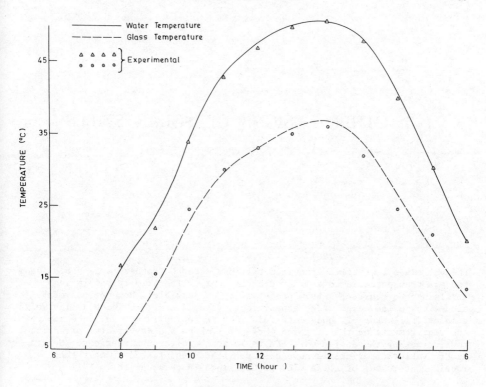

Fig. 5.16. Hourly variation of water and glass temperature on February 6, 1980 at Delhi, India. (After Sodha *et al.*, 1981a.)

6. OTHER DESIGNS OF SOLAR STILLS

The various solar stills discussed till now cater to rather large scale needs and are transportable to a limited extent only. The primary aim in such stills was to achieve as high productivity as possible. However, several situations arise, e.g. wars, expeditions, voyages etc., in which a lightweight compact still made of cheap materials is more appropriate. Here one may relax the requirements on the efficiency of performance and pay greater attention towards making the design more adaptable to the diverse situations. Several exotic varieties of stills have been fabricated and tested to meet these requirements; some of these will be described in this chapter.

6.1. LIFE RAFT TYPE SOLAR STILL

An all plastic, inflatable, floating type solar still for use along with a life raft in the oceans was designed for the U.S. Navy during the second world war by Maria Telkes (1945). The device, shown in Fig. 6.1, had no metallic or rigid parts, it could be folded into a small volume of about a litre and weighed about 1/2 Kg. Midway, inside a transparent plastic envelope was suspended a black porous pad, which could be soaked in sea-water. The porous pad absorbed solar energy resulting in evaporation of water. The vapors condensed on the inside surface of the plastic envelope and then trickled down into a collector bottle at the base of the still. So, one only had to inflate the still, get the pad soaked in sea-water and set the still afloat beside the raft. A ballast tank filled with sea-water was used to stabilize the floating still. The porous pad had to be flushed periodically in sea-water to get rid of the excess salt deposited in it.

A typical energy balance of this type of still is given in Table 6.1 (Telkes, 1945).

TABLE 6.1. Heat Losses and Efficiency for the Life Raft Type Solar Distiller

	Loss of solar energy, %
Transparent surface	8
Imperfact blackening	4
Reradiation (two sides)	20
Air circulation (two sides)	8
Total losses	40
Efficiency	60

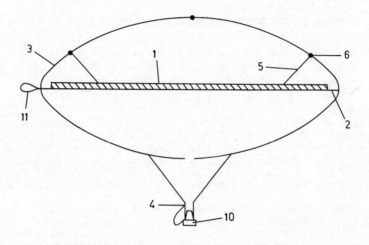

1 = BLACK POROUS PAD
2 = PAD SUPPORT
3 = TRANSPARENT ENVELOPE
4 = WATER-COLLECTING CHAMBER
5 = PAD SUSPENSION
6 = ATTACHING REINFORCEMENT
10 = PLUG
11 = TOWING LOOP

Fig. 6.1. Life raft type solar distiller. (After Telkes, 1945.)

A long tubular version of the raft type solar still was also proposed by Telkes (1945) for large scale production of distilled water. The major addition in this type was a saline water feeding system to supply the porous pad with sea-water at a slow continuous rate. This type of still could also be mounted on a frame.

6.2. FILM COVERED SOLAR STILL

A basin type still can be made very rugged and transportable by replacing the glass covers by plastic covers. Norov et al. (1975) suggested that an inflatable hemispherical plastic envelope can be used (see Fig. 6.2). This still also works on the same principle as the basin type still. The condensed water trickles down the sides of the envelope into a collector along the rim of the still.

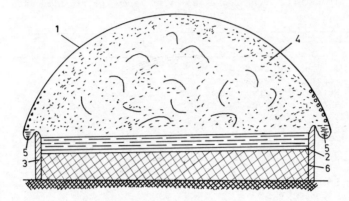

Fig. 6.2. Cross section of a film-covered solar still:
(1) transparent film; (2) black fabric;
(3) saline water; (4) vapor air mixture;
(5) trough for distillate; (6) thermal
insulation. (After Norov et al., 1975.)

A new variety of polyethylene sheets, based on polyvinyl alcohol, was prepared by Umarov et al. (1976). The effectiveness of these sheets could be increased by depositing a dispersion (trade name "Sun Clear") on them with a sprayer. It was observed that stills covered with the sheets treated with "Sun Clear" yielded 4.2 Kg/m^2 day of distilled water in July, 1975, at Tashkent, whereas the one with untreated film yielded only 2.7 Kg/m^2 day.

A serious problem with this type of still was, of course, the deterioration of the plastic cover. However, here too films treated with "Sun Clear" deteriorated less rapidly than the untreated films. For example, the transmission coefficient got reduced to 0.8 in the former case compared to 0.63 in the case of untreated plastic, after a use of four months.

6.3. WIPING SPHERICAL STILL (Menguy et al., 1980)

This type of still is similar in structure to the Raft-type still of Telkes et al. (1945), except that a blackened metal basin replaces the porous pad and an electrically operated wiper is attached to the top cover. The device as such is spherical. Water droplets, condensed on the interior surface of the upper hemisphere, are wiped by the wiper and directed into the lower hemisphere through the gap between the basin and the enclosing sphere (can be of glass). This system, developed by Menguy et al. (1980), is illustrated in Fig. 6.3.

It was found that wiping the inner surface of the cover enhanced the distilled water production by 25%. This may be attributed to the fact that the cover is maintained uniformly transparent to solar radiation and also, dripping of distilled water back into the basin is prevented.

Other Designs of Solar Stills 93

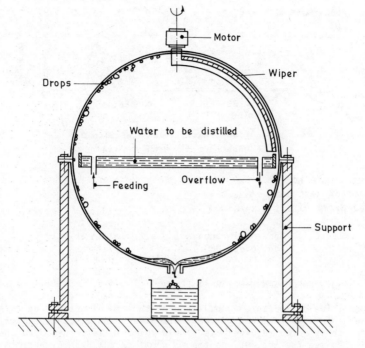

Fig. 6.3. Wiping spherical still. (After Menguy et al., 1980.)

6.4. CONCENTRIC TUBE SOLAR STILL

Falvey and Todd (1980) fabricated a new type of still (Fig. 6.4) using two concentric tubes. The larger tube, made of transparent flexible plastic, is closed at one end. The smaller tube, kept inside this tube, is made of metal and blackened to act as an absorbing surface; this inner tube is open at both ends. Saline water flows through a pipe wound along the entire length on the outside surface of the inner tube; the inlet and outlet for the saline water is at the same end of the still. Water oozes out of the pipe from perforations made on it.

The still works as follows. Air, at ambient temperature and humidity is blown through the annular space between the two tubes. The air becomes hotter because of the solar radiation absorbed by the inner tube. Consequently, it picks up moisture as the saline water oozes out of the piping. The warm humid air then flows into the interior of the inner tube and condenses on its cooler inner surface. The air is then exhausted into the atmosphere and the distilled water is collected. The latent heat given up at the walls of the inner tube preheats the saline water flowing in the perforated piping.

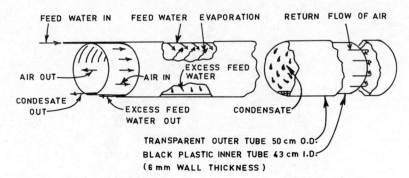

STILL LENGTH 60 m
AIR FLOW RATE 0.094 m³/sec.
FEED WATER RATE 0.74 Kg/(m·hr)

Fig. 6.4. Concentric tube solar still. (After Falvey and Todd, 1980.)

6.5. SOLAR EARTH-WATER STILLS

Large quantities of moisture accumulates in the ground during the rainy season, even in arid regions. Indeed, all plants and trees depend on this source for their water requirements. This moisture is returned to the atmosphere during the hot, dry months to complete the natural hydrological cycle. Several attempts have been made to exploit this for obtaining distilled water.

The earth-water still is similar to the basin type still with the ground replacing the basin. Solar energy, transmitted through the sloping glass cover, heats up the soil beneath the cover. As a consequence, vaporization of the moisture in the soil becomes significant. The vapors condense on the inner surface of the glass cover and the water collected in a collector-channel as usual. Earth water stills have been studied by Kobayashi (1963), and Ahmedzadeh (1978) in Iran. Ahmedzadeh observed no distilled water yield when the system was placed over ground. However, if the still, of area 1/2 m², was placed about 8cm and 28cm inside the ground, where the corresponding moisture contents were 9% and 11.2% respectively, 0.37 Kg and 0.5 Kg of distilled water could be collected after 24 hours.

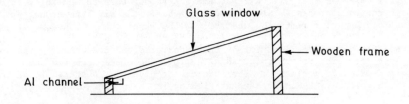

Fig. 6.5. Solar earth water still with wooden frame and glass window. (After Ahmadzadeh, 1978.)

In another interesting experiment Ahmadzadeh (1978) covered holes dug in the
ground with transparent plastic covers. This system was made leakproof by
means of soil itself (see Fig. 6.6). The cover was made concave, as seen
from outside, by placing a weight at its centre. A beaker was kept inside
the hole to collect fresh water. Moisture coming out of the ground condensed
on the cover and dripped into the beaker. The daily output from three stills
of this type, with soil moisture contents of 11.3%, 12% and 13.2% and with
hole diameters equal to 0.5m, is shown in Fig. 6.7. The output falls as the
number of days increase because the moisture content of the soil decreases.

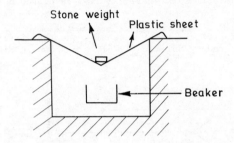

Fig. 6.6. Solar earth water still in the form of a hole
in the ground. (After Ahmadzadeh, 1978.)

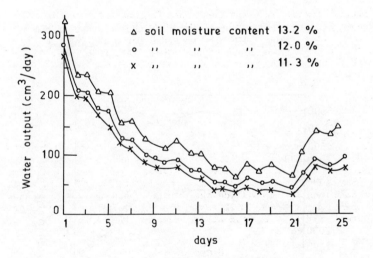

Fig. 6.7. Daily output from 3 holes in the ground-type
stills (30cm diameter) with soil moisture
contents of 11.3, 12.0 and 13.2%.
△ soil moisture content 13.2%
o soil moisture content 12.0%
x soil moisture content 11.3%
(After Ahmad-zadeh, 1978.)

This type of still may be an attractive emergency source of fresh water for
drinking in the arid regions.

6.6. COMBINED SOLAR COLLECTOR-BASIN TYPE STILL SYSTEMS

Soliman (1976) designed a composite system of a flat-plate collector and a basin type solar still (Fig. 6.8). The flat-plate collector of area 70×143 cm^2 was made using two corrugated aluminium sheets bonded together so as to form channels in which water can flow. It was provided with a glass top cover and a 5cm glass wool insulation below the absorber. Saline water heated in the flat-plate collector was conducted into a basin type still of area 59×170 cm^2. The still had a 3mm thick sheet of glass inclined at $30°$ to the horizontal as its cover. This still was also insulated below the basin by means of glass wool.

The diurnal variation of the solar intensity and the distilled water production, as measured by Soliman (1976), are shown in Fig. 6.9.

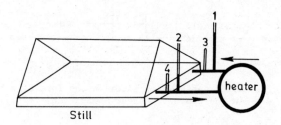

Fig. 6.8. 1,2. Tubes for measuring the pressure of the flow; 3,4. Thermometers for measuring the temperature of the flow. (After Soliman, 1976.)

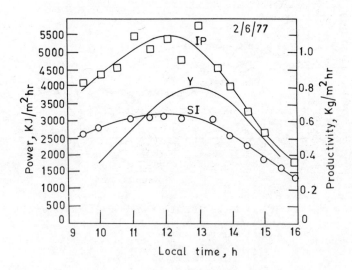

Fig. 6.9. Diurnal variation of solar intensity (SI), productivity (Y) and input power (IP) of the still. (After Soliman, 1976.)

6.7. AIR-SUPPORTED PLASTIC STILL (Lof, 1966)

This still consists principally of a shallow black basin or pond covered by a sloping transparent cover of glass or plastic. The pond contains saline water, which evaporates slowly on account of the solar energy absorbed at the base of the pond. The vapors condense on the cover, cooled by the external breeze, and flow into troughs kept at the rim of the pond or basin. This type of still differs from the conventional basin type still in that the water column is much deeper (typically 9m) and the construction is right on the ground. Heat losses into the ground from the base of the pond are also negligible. The principal virtue of the deep-basin still is its simple and cheap construction. On a typical sunny day, about 4 litre/m^2/day of distilled water is produced at a constant hourly rate.

In a slightly modified version, Delyannis (1965) maintained a higher air pressure inside the basin type still by means of a blower (Fig. 6.10). This automatically got inflated and hence supported the plastic cover on the basin. The system responded more rapidly to changes in solar radiation and ambient temperature. The water layer is shallower and hence the productivity is larger during the day time and very small during the nights. However, the net productivity during a 24 hour period is comparable to that of a deep basin still.

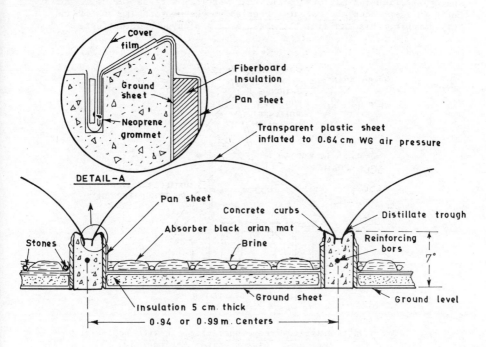

Fig. 6.10. Cross-section view of air supported plastic still. (After Lof, 1961b.)

6.8. EFFECT OF FLOATING MAT ON THE PRODUCTIVITY OF A STILL[*]

Very often a mat of loose plastic fibres is set afloat on the saline water in a still. The mat becomes wet due to capillary action in the pores and hence, presents a larger surface for evaporation. A new plastic material called Orlon, was developed out of poly-acrylonitrile for this purpose. Blackened non-woven mats made of Orlon fibres were found to withstand prolonged exposure to extreme conditions prevalent inside the stills; no brittleness or weakening were observed (Edlin, 1965). A sample removed after 11 months of use attested to the satisfactory quality of the material. Under normal operating conditions, the salts and other solid materials in the feed water should remain below the floating mats so that reflection from scale deposits is reduced to a minimum. Although initial experience with Orlon mat was not too encouraging, it must be stressed that during the 11 months of operation the sea-water feed was not screened or filtered nor was the operation of the still controlled.

6.9. SOLAR STILLS WITH REFLECTOR[**]

The use of reflecting surfaces, as shown in Fig. 6.11, has been limited to the applications of mirrors to the rear vertical surfaces of small units. While this enhances the output appreciably, the mirrors are costly and not very durable; this explains the absence of prototype applications.

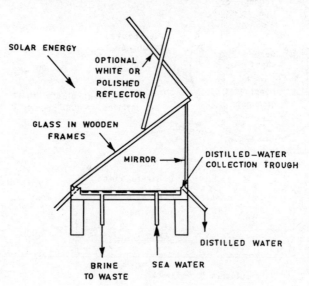

Fig. 6.11. Glass-covered evaporating pan with reflecting surfaces. (After United Nations, 1970.)

---

[*]After Lawand (1968).
[**]After United Nations, 1970.

Other Designs of Solar Stills

6.10. EXTRUDED PLASTIC STILL*

The sketch, of extruded plastic still with black wick for evaporation and cooling has been given in Fig. 6.12. Solar radiation heats the water in black wick after transmission through clear plastic extrusion; the water gets evaporated and condensed on the other side of the enclosure after giving its latent heat. This latent heat is lost to atmosphere through water flowing over it, and condensed water is collected in bottom of the enclosure.

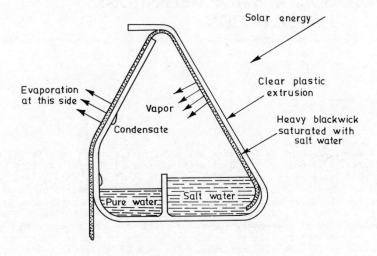

Fig. 6.12. Extruded plastic still with black wick for evaporation and cooling. (After United Nations, 1970.)

---

*After United Nations, 1970.

7. SOLAR STILL GREENHOUSE COMBINATION

With the development of greenhouses and water conservation techniques in agriculture, it is possible to have small-scale agricultural activity in places, where only saline or brackish water is available; solar distillation may in most cases be able to provide the rather modest demand for fresh water, consistent with clever choice of crops, suitable thermal environment provided by the greenhouse and the efficient water conservation technology. Thus an integrated design of greenhouse-cum-solar still presents an exciting possibility for support of small-scale agriculture in places where only saline or brackish water is available. The part of solar radiation, essential for the growth of plants may be provided by having part of the wall of glass or transparent plastic material or the base of the still, forming the roof of the greenhouse made of transparent material.

Several Greenhouses with solar stills incorporated in them, have been constructed and tested. A few of them are discussed in the following sections. The theoretical aspects are discussed in detail in section 7.3.

7.1. SOLAR STILL GREENHOUSE COMBINATION: THE TEXAS EXPERIENCE

Two small test greenhouses, with solar stills incorporated in them, were constructed in Texas by Qasim (1978).

The first unit (Unit No. 1; Fig. 7.1) was a two storey structure with a solar still in the upper part, the lower part being the greenhouse; the floor of the house was also used as a basin to hold water, so that the lower part also acted as a still. Plants were grown in pots and crates kept on the floor. The wooden frame comprising the house was covered with a 0.25mm clear polyethylene sheet. Some of the walls were made of Plexiglas sheets to provide rigidity to the structure and to enable entrances, ventilators and portholes (for probes) to be made. The roof was wedge shaped so that condensed water could trickle down along the roof into collector-troughs. The area of the basin (tough) of the upper still was $0.68m \times 0.15m$ and the trough was 0.08m deep. The base of the greenhouse was covered with a black plastic sheet. About 1.5 litres of brackish water was maintained in the lower basin.

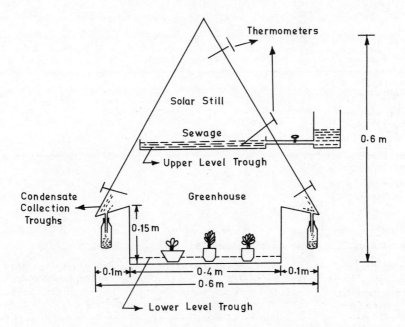

Fig. 7.1. Construction details of Unit No. 1. (After Qasim, 1978.)

The second unit (Unit No. 2; Fig. 7.2) was also constructed along the same lines as the first, the only alteration being that the top part containing the solar still was completely sealed-off from the greenhouse below. The entire house was covered with a 0.1mm thick plastic sheet instead of 0.25mm sheet used in Unit No. 1. The solar still dimensions were 0.58m × 1.45m × 0.10m. If required, the temperature inside the greenhouse could be reduced by painting the south and west facing walls white.

The performance of both these units is summarized in Table 7.1.

7.2. MATERIALS FOR GREENHOUSE SOLAR STILL SYSTEM

As in the case of the other stills, selection of cheap and durable materials for construction of still at the top of greenhouse is very essential if the system should become popular. The most important material is, again, the transparent cover used for the house. For letting in the solar radiation, essential for growth of plants, one uses transparent (to visible radiation) material for the base of the still; the amount of solar radiation, absorbed by the water and the base is utilized in the production of distilled water; this transparent material must preferably be commercially available and cheap. Thus, one has to separately study the photosynthesis (determining the growth) in plants and distillate output, corresponding to different transparent materials for the base of the still (i.e. roof of the greenhouse).

TABLE 7.1

Feature	Unit No. 1 February 10 to May 8, 75	Unit No. 2 July 8 to September 1, 75
Highest temperature in solar still (°C)	61	63
Daytime average temperature in solar still (°C)	50.6	51.1
Sewage temperature in evaporating pan (°C)	31.1	Slightly lower than 51.1
Temperature in the greenhouse (°C)	43.9	42.8 (door was closed)
Condensate produced	0.33 Kg/day	0.23 Kg/day
Condensate quality	average pH=7.2	pH=8.0 to 8.7
Plant viability	None of plants survived	All plants survived in the greenhouse
Relative humidity in greenhouse	0.90 (doors were closed) 0.38 (doors were opened)	

Tinaut et al. (1978) constructed a greenhouse, shown in Fig. 7.3, to study the viability of several kinds of polyethylene sheets. The base of the solar still, lodged in the upper part of the house, was made of semi-transparent plastic which permitted sufficient amount of visible solar radiation to penetrate into the greenhouse; this helps plant growth due to improved photosynthetic activity.

The absorption spectrum of chlorophylls (maximum between 640-670nm and 410-440nm) and carotenoids (maximum between 400-450nm) is modified only slightly when these pigments are combined with other molecules. The fact that there is no commercial plastic material with absorption spectrum in this range allows a free choice. Tinaut et al. (1978) selected the following:

1. Red methacrylate   rigid
2. Red polyethylene   flexible
3. Blue methacrylate   rigid
4. Blue polyethylene I   flexible
5. Blue polyethylene II   flexible (more opaque than other one)

7.2.1. Measurement of Photosynthetic Activity

The photosynthetic activity is expressed in fixed micromols of carbon dioxide per square centimeter of leaf area per hour of exposure.

Solar Still Green House Combination

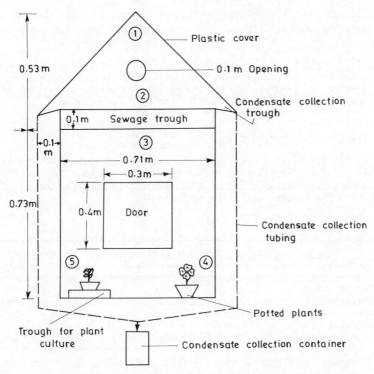

Fig. 7.2. Construction details of Unit No. 2. (After Qasim, 1978.)

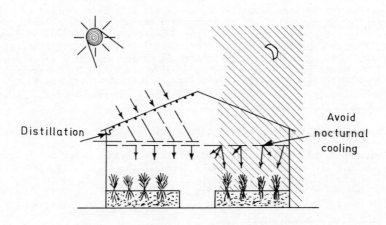

Fig. 7.3. (After Tinaut et al., 1978.)

Experiments were conducted on freshly germinated barley plants to determine the rate of photosynthesis due to radiation passing through the transparent material under test. A sample of barley leaf, 14mm in diameter, was prepared and kept in the test chamber. It was then irradiated with light passing through the test material, while controlled amounts of air, containing CO_2 gas labelled with radio-active isotope of carbon C^{14}, was pumped into the chamber. The extent of photosynthetic activity was determined by monitoring the amount of radio-activity in the barley leaf.

Five barley plants were utilized for testing each material. It was concluded that all the selected plastic materials were acceptable. Further, maximum photosynthetic activity was obtained with the red methacrylate and minimum photosynthetic activity corresponded to the blue polyethylene.

7.2.2. *Estimation of Still Production*

Tinaut et al. (1978) also considered the spectral distribution of solar radiation and the absorption curves of these plastics to calculate the absorbed radiation for each material and hence predicted the output of the distillate. The results are given in Table 7.2.

TABLE 7.2

Material	Relative photo-synthetic activity %	Annual water production Kg/100m^2
Red methacrylate	102.6	35.05
Red polyethylene	69.5	21.06
Blue methacrylate	52.7	23.81
Blue polyethylene I	44.3	14.35
Blue polyethylene II	55.7	25.02

7.3. STILL ON ROOF: ANALYSIS

7.3.1. *Background*

Although several authors have studied in detail different aspects of solar distillation and thermal load levelling in buildings, little attention has been paid to the use of solar stills for controlling environment along with distilled water production; the work has been restricted to greenhouses. Selcuk (1970, 1971) has analyzed the thermal performance of a greenhouse (Fig. 7.4), whose roof is covered with a solar still; the computer program, apart from being rather complex, is difficult to use, is limited to calculations concerning greenhouse only and does not take thermal conduction in the ground into account properly.

Sodha et al. (1980b) have proposed that a still can be placed on the roof of a building for producing distilled water and, at the same time, assisting in air conditioning of the building. Analytical expressions for the hourly heat flux into the room and the hourly yield of the distillate were obtained using linearized Dunkle's relations (section 3.4). Numerical calculations were carried out for typical cold (9th March 1979) and hot (19th June 1979) days in Delhi. The effect of dye in the saline water and of absorptivity of the bottom surface (when the dye is not used) on the performance of the system has also been studied for typical cold and hot days respectively. It is interesting to note that this system provides distillate water in addition to the control of thermal load in the building. It is concluded that:

(i) for a typical hot day (19th June 1979), there is a reduction of heat flux by 40% in addition to distillate production of 0.6 Kg/m^2 day (when the dye is not used); the τ_3 of the bottom surface of the basin has been taken as 0.2.

(ii) for a typical cold day (9th March, 1979), there is an enhancement of heat flux by a factor of two with distilled water yield of 5 Kg/m^2 day (when a dye is used). The yield of distillate in this case is larger than that corresponding to the conventional still because the heat loss is lower on account of the insulating effect of concrete and the temperature of air in the room being higher than the ambient temperature.

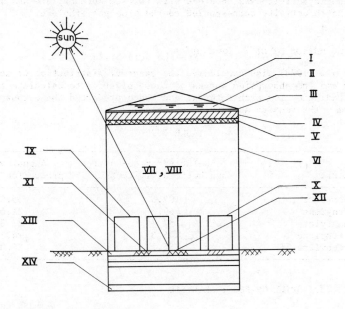

I   Solar still glass cover
II Brine
III Basin liner
IV Insulation
V   Solar still bottom
VI Greenhouse glass cover
VII Air stream control volume for heat balance
VIII Air stream control volume for mass balance
IX Sunlit plant canopy
X   Shaded plant canopy
XI Sunlit soil
XII Shaded soil
XIII } Ten soil slabs
XIV

Fig. 7.4. Mathematical model of the solar still greenhouse. (After Selcuk, 1970.)

7.3.2. Analysis

The proposed system is shown in Fig. 7.5. Following Sodha *et al.* (1980b) the energy balance conditions for glass cover, basin water and basin liner can be written as

$$M_g \frac{dT_g}{dt} = \tau_1 H_s + (q_{r\omega} + q_{c\omega} + q_{e\omega}) - q_a \quad , \tag{7.3.1}$$

$$M_{\omega o} \frac{dT_\omega}{dt} = \tau_2 H_s + q_\omega - (q_{r\omega} + q_{c\omega} + q_{e\omega}) \quad , \tag{7.3.2}$$

$$\tau_3 H_s = q_\omega + q_{ins} \quad . \tag{7.3.3}$$

Here, $q_{c\omega}$ and $q_{e\omega}$ may be written with the help of Eqn. (3.4.4) as

$$q_{c\omega} = h_{c\omega} (T_\omega - T_g) \tag{7.3.4}$$

and

$$q_{e\omega} = h_{eff} (T_\omega - T_g) \tag{7.3.5}$$

where $h_{c\omega}$ and h_{eff} are same as in section 3.

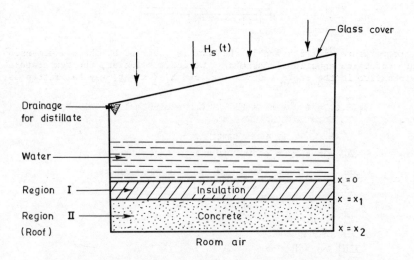

Fig. 7.5. Schematic sketch of "still on roof" system. (After Sodha *et al.*, 1980b.)

Energy transferred from the absorbing surface to water in the basin can be written as,

$$q_\omega = h_3 \left[\theta_1 \big|_{x=o} - T_\omega \right] \qquad (7.3.6)$$

Equations (7.3.1)-(7.3.3) may now be rewritten as,

$$M_g \frac{dT_g}{dt} = \tau_1 H_s + h_1(T_\omega - T_g) - h_2(T_g - T_a) \quad , \qquad (7.3.7)$$

$$M_{\omega o} \frac{dT_\omega}{dt} = \tau_2 H_s + h_3(\theta_1\big|_{x=o} - T_\omega) - h_1(T_\omega - T_g) \quad , \qquad (7.3.8)$$

$$\tau_3 H_s = h_3 (\theta_1\big|_{x=o} - T_\omega) - K_1 \frac{\partial \theta_1}{\partial x}\bigg|_{x=o} \qquad (7.3.9)$$

The value of h_3 is different when the heat flows upward i.e. $\theta_1\big|_{x=o} > T_\omega$ compared to that when it flows downwards, i.e. $\theta_1\big|_{x=0} < T_\omega$. Considering the temperature and heat flux to be continuous at the surface $x = x_1$, the boundary conditions at $x = x_1$ may be written as:

$$\theta_1\big|_{x=x_1} = \theta_2\big|_{x=x_1} \qquad (7.3.10)$$

$$-K_1 \frac{\partial \theta_1}{\partial x}\bigg|_{x=x_1} = -K_2 \frac{\partial \theta_2}{\partial x}\bigg|_{x=x_1} \qquad (7.3.11)$$

Since the plane $x = x_2$ is in contact with the room air at constant temperature, the energy balance may be expressed as:

$$-K_2 \frac{\partial \theta_2}{\partial x}\bigg|_{x=x_2} = h_4 \left[\theta_2\big|_{x=x_2} - \theta_R \right] \qquad (7.3.12)$$

The temperature distribution in the roof is governed by the one dimensional heat conduction equation. Assuming a periodic solution, the temperature distribution in the regions $0 < x < x_1$ and $x_1 < x < x_2$ may be written as,

$$\theta_1(x,t) = A_1 x + B_1 + \mathrm{Re} \sum_{n=1}^{\infty} \{C_{1n} \exp(-\beta_{1n} x)$$

$$+ D_{1n} \exp(\beta_{1n} x)\} \exp(in\omega t) \quad , \qquad (7.3.13(a))$$

$$\theta_2(x,t) = A_2 x + B_2 + \mathrm{Re} \sum_{n=1}^{\infty} \{C_{2n} \exp(-\beta_{2n} x)$$

$$+ D_{2n} \exp(\beta_{2n} x)\} \exp(in\omega t) \qquad (7.3.13(b))$$

where,

$$\beta_{1n} = -\alpha_1\sqrt{n}\,(1+i)\quad,\quad \alpha_1 = \left(\frac{\omega\rho_1 C_1}{2K_1}\right)^{1/2},$$

$$\beta_{2n} = -\alpha_2\sqrt{n}\,(1+i)\quad,\quad \alpha_2 = \left(\frac{\omega\rho_2 C_2}{2K_2}\right)^{1/2}.$$

The solar intensity and ambient temperature can also be assumed to be periodic.

From the nature of the basic differential equations, it can be inferred that T_g and T_ω are periodic. Thus,

$$T_g = g_o + \mathrm{Re}\sum_{n=1}^{\infty} g_n \exp(in\omega t)\quad, \qquad (7.3.14(a))$$

$$T_\omega = H_o + \mathrm{Re}\sum_{n=1}^{\infty} H_n \exp(in\omega t) \qquad (7.3.14(b))$$

The constants A_1, A_2, B_1, B_2, H_o, g_o, C_{1n}, C_{2n}, D_{1n}, D_{2n}, g_n and H_n may be determined by the substitution of Eqns (7.3.13) and (7.3.14) in Eqns (7.3.7)-(7.3.12).

The heat flux through the roof is given by

$$\dot{Q}(t) = h_4(\theta|_{x=x_2} - \theta_R)$$

$$= h_4\left[A_2 x_2 + B_2 - \theta_R + \sum_{n=1}^{\infty}\{C_{2n}\exp(-\beta_{2n}x_2)\right.$$
$$\left. + D_{2n}\exp(\beta_{2n}x_2)\}\exp(in\omega t)\right] \qquad (7.3.15)$$

The heat flux per unit area, associated with evaporation is

$$q_{e\omega} = \mathcal{L}\cdot\dot{m}_e$$
$$= h_{eff}(T_\omega - T_g)$$
$$= h_{eff}[(H_o - g_o) + \sum_n (H_n - g_n)\exp(in\omega t)] \qquad (7.3.16)$$

The mass of water evaporating M_e, per day in Kg/m^2 can be obtained by integrating the R.H.S. of Eqn. (7.3.16) over 24 hours (i.e. one full period); thus

$$M_e = \int_0^{24\times 60\times 60} h_{eff}\cdot\frac{(T_\omega - T_g)}{\mathcal{L}}\,dt$$

$$= \frac{h_{eff}(H_o - g_o)}{\mathcal{L}}\times 24\times 60\times 60\quad Kg/m^2\text{ day}$$

7.3.3. *Numerical Calculations and Discussion*

To have a numerical appreciation of the analytical results, calculations for the heat flux into the room and the hourly yield of distillate have been carried out using the hourly data of solar radiation and atmospheric temperature for a typical (i) hot day (19th June, 1979), and (ii) cold day (9th March, 1979) in Delhi. On the basis of results thus obtained, the following conclusions were arrived at:

1. When $\tau_3 = 0.65$ (referring to a black bottom) the average heat flux is large but the swings are small corresponding to a hot day. In due course of time, the deposition of wastes on the bottom should result in a lower value of τ_3 and subsequent reduction of heat flux into the room. The value $\tau_3 = 0.2$ corresponds to a reduction in heat flux entering the room by 40% compared to a bare roof and a distillate output of 0.6 Kg/m^2.

2. The productivity of distillate decreases significantly with decreasing τ_3.

3. Thermal insulation over the concrete base of the still improves its performance to some extent, but excessive insulation has the opposite effect. However, insulation on either side of concrete makes only marginal difference on the time dependence of heat flux.

4. The corresponding effect of insulation on the productivity of the still has also been studied. Obviously, owing to reduced conduction losses the output increases with increasing insulation thickness.

5. In winter, the presence of a dye in the basin water results in increased collection of solar energy; this causes an increase in the heat flux entering the room (compared to a bare surface) and the productivity of the still increases as well. It may be noted here that for cold climates, the insulation may not be necessary.

8. ECONOMIC ASPECTS OF SOLAR DISTILLATION

In the previous chapters we have discussed the technical aspects of solar stills. However, the utilization of solar stills as a source of fresh water for drinking, industry and agriculture on a small scale will be essentially determined by the economic viability. A still, to be economically justifiable should, within its life span, pay back more than the entire money invested on it and be preferably the least capital intensive of the alternatives for fresh water supply in a given region.

Distillation of brackish/sea water for supplying fresh water is thought of, obviously, only for those regions that do not have a nearby supply of natural fresh water. The main techniques for distillation, which are technically viable, are (a) flash distillation, (b) vapor compression process, (c) electrodialysis, (d) reverse osmosis and (e) solar distillation. Depending on the existing conditions in a particular region one of these may be more economical than others. In making this decision, one is guided by the following considerations:

1. Quantity of fresh water required and its end use.
2. Available water sources, such as sea, ponds, wells, swamps etc.
3. Proximity to nearest fresh water sources.
4. Availability of electric power at the site or closeby.
5. Cost of supplying fresh water by the various methods.
6. Cost and availability of labor in the region.
7. Maintenance and daily operational requirements.
8. Life span of the water supply system.
9. Economic value of the region.

When the demand for fresh water is small, say about 10,000 gallons* per day (gpd) as in desert pasture lands, it is highly uneconomical to instal pipe lines to supply fresh water to that region. At times fresh water was supplied by railway tankers, trucks, mules and even human beings at costs as high as $7 per 1000 gallons (United Nations, 1970). By employing solar distillation techniques, water could be supplied at less than half of this price (United Nations, 1970). The deciding factors would, of course, be the

---

*1 gallon = 3.78 litres.

economic value of the region and its distance from natural fresh water sources. For instance, in the arid zones of the Turkeman SSR, it was found more economical to instal solar stills at locations farther than 80 kms from fresh water sources than to supply it by trucks or by collection of rainfall (Baum and Bairamov, 1968). In the central and eastern parts of Kara-Kum deserts in USSR, the cost of water brought by trucks over a distance of 70 km was $20 to $28 per 1000 gallons (Baum et al., 1967). On the Makram coast of Pakistan the cost of water brought by trucks is about $48 per 1000 gallons (United Nations, 1970). Numerous other examples of high water costs, at small levels of demand, can be cited.

It must be stressed that these cost projections of early 1960's are no longer valid because of the big inflation witnessed since then and, particularly, due to the ten-fold rise in petrol prices. Indeed, these factors have made solar energy more competitive now, since the cost of transport (because of use of diesel oil) has gone up much more than that of labor and other materials, which are used in the fabrication of stills.

In many countries, there are vast grasslands and semi-deserts which have known supplies of saline water or shallow brackish aquifers. In such areas, with a modest supply of fresh water for drinking, sheep and similar livestock may be profitably reared. Solar stills fed with saline water from wells by wind operated pumps (and completely automated, except for periodic inspections) have been designed for this purpose in several countries. An Australian estimate (United Nations, 1970) shows that a solar still of about 420m^2 area can be built for about $2,500 to supply water to a stock of 350 sheep at a cost of $4 per 1000 gallons; for a net investment of about $4,500 the returns were better than 15% (United Nations, 1970). Estimates based on studies at Turkeman SSR have led to similar conclusions (United Nations, 1970). It thus appears that, solar stills requiring only occasional attention should be of considerable use for economically exploiting hitherto unused pastures and semi-deserts. A series of appropriately distributed solar desalination plants could be an ideal solution to this problem. One caution must, however, be exercised, viz. the disposal of concentrated brine discharged from the still should be carefully planned.

8.1. COST OF PRODUCT WATER*

It is clear from the above discussions that the crux of the matter is the cost of fresh water available at any place. For a distillation unit the major items of cost per annum would be

(a) the total fixed costs as an annual percentage of capital investment,

(b) cost of supplying saline water to the distiller, and

(c) operating labor and supervision expenses.

The sum of these expenses divided by the annual output of distilled water from the still plus assured collection of rainwater would be the cost of production of the distillation plant. If the entire quantity of fresh water produced is gainfully utilized, then this would also be the cost of water

---

*After United Nations (1970).

supplied; otherwise, loss due to wastage would also get reflected in the cost of water supplied. Thus, if $\overline{IA}$, $\overline{MR}$ and $\overline{TI}$ represent the average value over the estimated life span of the installation, the cost of fresh water is given by

$$C = \frac{10I(\overline{IA} + \overline{MR} + \overline{TI}) + 1000\,(O \cdot C' + S)}{A(Y_D + Y_R)}$$

where

C = cost of water (dollars per 1000 gallons);

I = total capital investment (dollars);

$\overline{IA}$ = annual interest and amortization rate (percentage of investment);

$\overline{MR}$ = annual maintenance and repair labor and materials (percentage of investment);

$\overline{TI}$ = annual taxes and insurance charges (percentage of investment);

O = annual operating labor (man hours);

C' = operating labor wage (dollars per man hour);

Y_D = annual unit yield of distilled water (gallons/m^2);

Y_R = annual unit yield of collected rainwater (gallons/m^2);

A = area of distiller on which yields are based (m^2);

S = total cost (fixed and operating) of salt-water supply.

8.2. EARLY ECONOMIC ANALYSIS*

The first significant economic analysis of solar distillation systems was made by Lof (1962). In this study the cost of production of distilled water from sea water by solar distillation was compared with that by vapor compression and multiple-stage flash distillation techniques; the study was limited to plant capacities up to 100,000 gpd. Solar still construction costs were obtained from previous estimates and the prices of the other two systems from manufacturers. Operating costs were worked out from estimates of labor requirements and the manufacturers data. An optimistic life span of 50 years was assumed for the asphalt-glass solar still.

The results of the analysis are summarized in Fig. 8.1. For the sake of comparison costs of water were also correlated with plant size on a 20-year amortization basis. It is seen that for plant capacities less than 50,000 gpd, solar distillation is the cheapest among the three techniques. By completely automating the operation of a solar still, which is quite possible, significant reduction in the cost of distilled water may be achieved as indicated by the curves labelled "no operating labor". However, these cost projections are largely approximate on account of the numerous assumptions made regarding labor requirements, wage rates, depreciation rates etc.

---

*After Lof (1966).

Economic Aspects of Solar Distillation

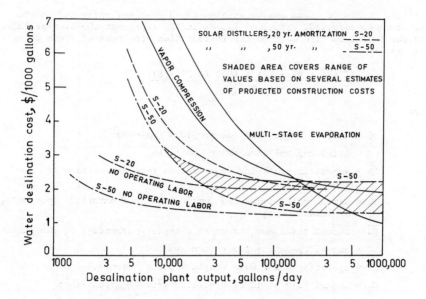

Fig. 8.1. Costs of water from solar and conventional distillers at various plant capacities. (Lof, 1962.)

Similar conclusions have been arrived at by Bloemer *et al.* (1965a) in their cost analysis based on solar energy-supply rate of 23,000 KJ/m^2 day and labor cost at $0.50 per man hour, typical of developing countries. A 20-year amortization period at 4% annual interest was assumed for all plants. The results are shown in Fig. 8.2, from which it is seen that distilled water from a 50,000 gpd plant would cost $2.30 per 1000 gallons and that from a 10,000 gpd plant would cost $2.70 per 1000 gallons. The curve from Fig. 8.1 corresponding to solar still amortized for 20 years is also shown in Fig. 8.2 for comparison. It is seen that the agreement between the two studies is fairly good. The difference in the slopes of the curves may be ascribed to the assumption of lower labor cost and higher plant amortization in the studies due to Lof (1962).

In an earlier study, Bloemer *et al.* (1964) correlated the cost of production of distilled water to the cost of the plant, its working life span and its efficiency; this is summarized in Fig. 8.3. It is seen that a permanent type of distiller made of glass and asphalt, costing $10/m^2 and operating at 45% efficiency with a service life of 30 years, would yield distilled water at $1.75 per 1000 gallons. To give the same performance, a plastic-sheet distiller, operating at an efficiency of 60% with a service life of 5 years, should cost less than $3.8/m^2. Here, the amortization charges were computed at 4% annual interest. If the life span of the plastic still could be raised from 5 to 10 years, a manufacturing cost of $6.46/m^2 can be tolerated. However, if the still were to be amortized in as little as 3 years, it would have to be fabricated at a cost of less than $0.82/m^2 and with very little labor.

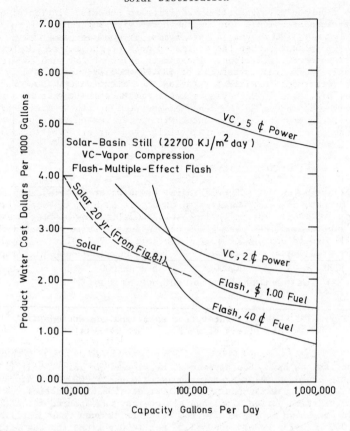

Fig. 8.2. Costs of water from solar and conventional distillers at various plant capacities. (Bloemer *et al.*, 1965.)

8.3. RUSSIAN EXPERIENCE*

The Russians propose to raise a stock of 260 thousand sheep in the Erbent State Farm in the desert regions of Ashkhabad. The capital investment required for laying the Erbent water line to supply fresh water to the farm, of area 3.14 million hectares, is estimated to be $79.32 million. On the other hand, construction of 260 appropriately distributed solar desalination plants would be sufficient to meet the requirements of the farm. The cost of construction of one such plant would vary from 66.69 to 111.24 thousand dollars, depending on the depth of the wells, salinity of the water and distance from the factory manufacturing the stills. Normally, wells were found to be 20m deep with salinity of the order of 15 to 25 gms/litre. The estimated cost of a solar desalination plant, at a distance of 100 km from the factory and supplied by a well 50m deep with water of salinity 25 gms/litre

---

*After Bairamov *et al.* (1979).

Economic Aspects of Solar Distillation   115

is 88.14 thousand dollars. If in addition to the distillation plant other facilities such as a house for the shepherd, with solar heating, cooling and hot water supply, sheep pens, water lifter, greenhouse for growing vegetables etc. are also planned, then the estimated cost of the entire complex would come up to about $0.14 million. The construction of 260 such complexes would require a capital investment of $35.75 million, which is only 45% of the total investment required for laying the Erbent water line. Further, while the solar desalination plants would pay back the entire investment in 4 to 7 years, depending on the depth of the well and the salinity of the well water; the proposed Erbent water line is estimated to require 18 years for the pay off. A more eloquent justification for erecting solar desalination plant is hard to come by.

8.4. INDIAN EXPERIENCE

In the marshy regions of Kutch in India, fresh water is being supplied by means of trucks, tankers, camels etc. at a cost as high as $5 per 1000 gallons (Datta *et al.*, 1965). By installing solar desalination plants, fresh water could be supplied at less than $1000 per gallons. It is estimated that a solar still should cost about $50/m^2 (Gomkale, 1980). Thus, solar distillation should be an attractive alternative to existing schemes of water supply in the Kutch.

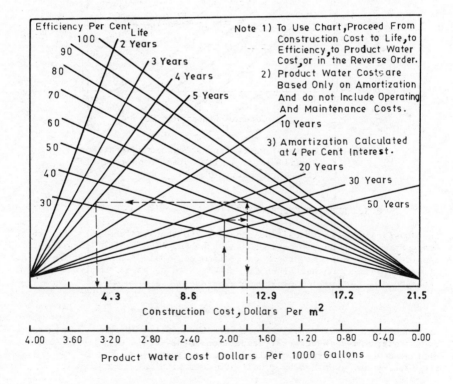

Fig. 8.3. Cost comparison chart for solar distillation. (Lof, 1966.)

Gomkale (1969) made a detailed economic study of the comparative merits of a number of schemes for supplying fresh water in the desert regions of Rajasthan in India. His studies are summarised in Figs. 8.4 and 8.5. The solid curves in Fig. 8.4 correspond to the costs of pipeline systems of varying lengths; in these calculations the cost of operation, maintenance and labor have not been taken into account. It is seen that for a pipeline 30 km long it is more economical to instal solar stills if the water requirements are about 4000 gallons/day.

Figure 8.6 shows the relative performance of a solar still and a Humidification-Dehumidification (H-D) plant, proposed by Garg *et al.* (1966). From this figure also it is clear that for water requirements less than 20,000 gpd it is better to install a solar desalination plant.

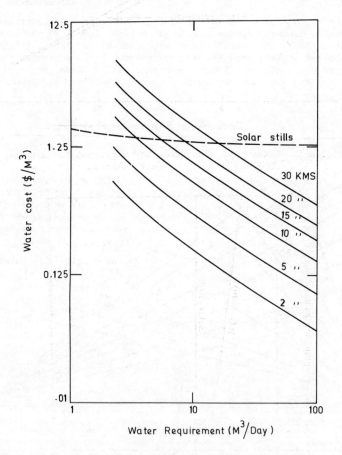

Fig. 8.4. Water cost comparison solar stills and G.I. piping. (After Gomkale, 1969.)

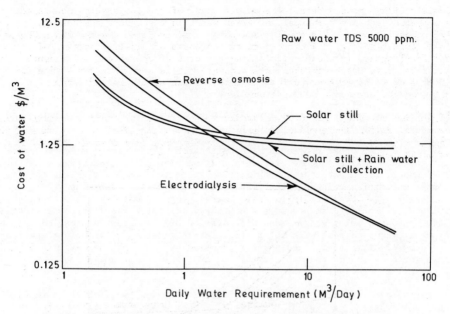

Fig. 8.5. Water cost comparison for different techniques of desalination. (After Gomkale, 1969.)

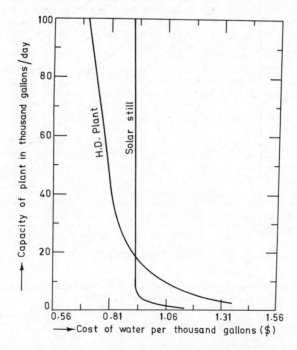

Fig. 8.6. Cost of water/1000 gallons for solar still plant and for humidification-dehumidification plant. (After Garg *et al.*, 1966.)

8.5. SOLAR DISTILLATION VERSUS CONVENTIONAL PROCESSES[*]

From the discussions made so far it is clear that the cost of production of distilled water from solar stills is about $3 per 1000 gallons for plants of capacities larger than 5000 gpd. The cost would rise at lower capacities because of labor wages at the manufacturing plant do not get reduced proportionately. The labor charges remain more or less the same (i.e. $3 per 1000 gallons) for plants of larger capacities, wherein the economy of scale operates; the capital cost of installation per unit capacity would not decrease as rapidly as the cost of process equipment. Upto plant capacities of 50,000 gpd, the cost of fresh water from solar stills remains the lowest, but beyond this other conventional distillation plants become more economical. For instance existing commercial desalination plants, of capacities 1 million gpd, supply fresh water at approximately $1 per 1000 gallons. Plants of higher capacities being designed may reduce this cost to less than $0.50 per 1000 gallons. Solar stills in their current form cannot compete with these large distillation plants. On the contrary, for water requirements less than 50,000 gpd the multi-stage flash distillation or the vapor compression process cannot compete with solar distillation. Perhaps, the only competitor could be electrodialysis and reverse osmosis, but the cost of water in this case would depend on the salinity of the available water and also the availability of electric power.

It may be concluded that if a need exists in a community in which there is abundant sunshine and in which there is a requirement for potable water in the range of up to, say, 25,000 gpd, a solar still can provide the most economical supply of desalted sea water. For demands appreciably above this level, further analysis of factors must be made, including variation in seasonal demand, variation in seasonal output of solar distiller, specific cost factors, availability of storage and most recent information on the cost of conventional processes in the six ranges required. Above about 50,000 gpd, non-solar methods appear to be most economical.

Another point worth noting is that for rapidly expanding water needs, the solar distillation plants may not be the ideal choice. However, within the range in which they are economical, the modular structure of solar stills may be used with advantage to expand existing plant capacities; note that there is no need for over-designing the plant initially.

The constructional costs and other expenses involved in several major solar stills have been summarized in Table 8.1. It is seen that labor costs are high in the United States (35% of the total cost of the still per unit area), whereas in Greece they are low (19% of the total cost of the still per unit area). The material costs are almost the same. Thus, unless large scale import of construction material is required, the economics of solar stills will be governed by the labor wage rates existing in the region.

---

[*]After United Nations (1970).

TABLE 8.1. Capital Costs of Solar Stills (After United Nations, 1970)

	Las Marinas still (9350 sq.ft.) 868m² 1966	Daytona Beach still (2650 sq.ft.) 246m² 1961			Patmos still (86000 sq.ft.) 8000m² 1967			Projected still USA. 9300m² (1 million sq.ft.)		
	Actual costs (dollars per sq.ft.)*	Actual costs (dollars per sq.ft.)*			Actual costs (dollars per sq.ft.)*			Actual costs (dollars per sq.ft.)*		
		Materials	Labor	Total	Materials	Labor	Total	Materials	Labor	Total
Land preparation	0.026	0.04	0.32	0.36				0.01	0.09	0.10
Concrete beams	0.211	0.25	0.15	0.40				0.19	0.10	0.29
Masonry	0.063	0.07	0.04	0.11	0.09			0.16	0.04	0.10
Basin lining	0.502	0.32	0.19	0.51	0.26			0.15	0.10	0.25
Glass covers	0.502	0.37	0.14	0.51	0.23			0.25	0.10	0.35
Piping channels	0.188	0.03	0.01	0.04	0.06			0.03	0.01	0.04
Miscellaneous equipment	0.246				0.05					
Electrical	0.178				0.05					
Special works	0.146				0.19					
Distillate troughs		0.03	0.02	0.05				0.03	0.02	0.05
Aluminium frames					0.28					
Contractor's fee				0.32						0.12
Subtotal		1.11	0.87		1.21			0.72	0.46	
Total	1.56			2.30			1.51			1.30

*1 ft² = 0.029m²

It may be noted that most of the analyses cited above were performed before 1970. However, the world has seen, since then, a tremendous rise in prices of raw materials, cost of labor and general inflationary rise. Consequently, all the economic projections of early 1970s have become obsolete. In fact the tenfold hike in cost of petroleum has been the single most important factor responsible for this. Thus, what was economically not viable in the early 1970s has become viable now, which goes in favour of solar technology. The general conclusions drawn in the earlier economic studies acquire a stronger basis. It should, however, be asserted that a proper appraisal based on the current economic situation should be undertaken at the earliest to put the entire subject in proper perspective.

9. RECOMMENDATIONS FOR FUTURE RESEARCH

Many analyses of single basin solar stills have been proposed and validated; some analyses of double basin solar stills and multiple wick analyses are also available. However, no analysis for a number of designs of solar still appears to have been made although some empirical correlations of limited validity have been made. Hence there is a distinct need for developing analytical models for different types of stills, which may be validated experimentally and an optimum design evolved; the performance of the stills over long periods and the economics also need to be studied.

The heat and mass transfer phenomena in the stills has also not received enough attention. The presently used relations are valid in a limited range of Grashof number and equal horizontal areas of evaporating and condensing surfaces. The generalization of these results for application to the different designs of the stills is also necessary.

Stills in which the glass cover is cooled by flow of water above it also need to be investigated; the effect of forced convection and partial vacuum inside the still is also worth investigating.

Studies on performance and economics of solar desalination are based on the use of single basin stills and are outdated because of the increase in the cost of materials, labor and fuel. Such studies should be extended to include other designs and present cost parameters. Economics of use of solar distilled water in small scale industry, laboratories and hospitals should also be worked out.

10. APPENDIX A

SOLAR STILL PLANT AT AWANIA VILLAGE, GUJARAT, INDIA
S.D. GOMKALE, CENTRAL SALT AND MARINE CHEMICAL RESEARCH
INSTITUTE, BHAVNAGER, 364002, INDIA

1. Location:

   Awania is a non-electrified village located at about 12 Kms away from Bhavnagar on Bhavnagar-Gogha road, in Gujarat, India.

2. Climatological data:

   The required data for Bhavnagar is given in Table A-1. At Awania we have not placed instruments except for an anemometer.

   TABLE A-1. Climatological and Solar Radiation Data of Bhavnagar

Month	Temperature Average maximum in °C (last 10 years)	Temperature Average minimum in °C (last 10 years)	Wind velocity in Kms/hrs (last 10 years)	Actual sunshine hours (last 10 years)	Total solar radiation in horizontal surface (last 13 years) (in cals/cm² /day)
January	28.2	13.0	7.31	304.5	429
February	30.3	15.3	8.28	290.5	495
March	35.2	20.3	8.84	316.8	574
April	38.3	24.4	10.02	322.1	614
May	39.7	26.1	12.07	353.7	636
June	36.1	26.4	12.63	199.7	480
July	33.1	25.6	11.48	96.0	383
August	32.2	24.6	10.96	91.1	347
September	33.2	23.7	8.96	197.0	482
October	35.2	23.1	6.92	294.5	488
November	32.2	18.5	6.56	275.4	399
December	29.0	14.2	6.60	298.0	396
Total	402.7	255.2	110.63	3039.3	5726
Average	33.6	21.3	9.22	253.3	477

3. Salt concentration of the underground water at Awania Village:

   The total dissolved salts (TDS) vary in the well water throughout the year with maximum value during summer months as seen in the following data. The fluoride concentration is above 6 ppm.

Date of sampling	TDS, ppm
September 1976	3010
17.3.77	4000
27.6.78	5039
4.10.78	4128
24.1.79	5002
20.3.79	6512
16.4.80	5861
20.4.81	6130

Appendix A   123

4. Village Statistics:

   Population: 1413
   Cattle population: 671

5. Description of the plant:

   The plant has a mean capacity of 5000 litres/day. It has an evaporating
   surface of 1866.6 sq.m. distributed over 90 stills which are grouped in
   15 blocks of 6 stills each. The layout of the plant is shown in
   Fig. A-1. In December, 1977 the plant was commissioned; a pictorial
   view can be seen in the photograph A-2.

   Individual stills are 12.2m long and 1.7m in width and the entire
   construction is in brick masonry. A water depth of 7 to 10cm is
   maintained inside the still. Ordinary window pane glass of 3mm thickness
   is used as glass cover which is fixed at an angle of 20°. Top edges of
   glass sheets rest on an aluminium tee fixed on supporting pillars whereas
   lower edges rest on aluminium distillate channels fixed on the walls.
   Slope for drainage of distilled water is provided in the channel itself.
   Product water from all the stills is collected in an underground tank.
   This water is pumped into the supply tank for villagers, who collect it
   for their use. The stills are operated batchwise with saline water fed
   by gravity from the feeding tank constructed above ground level.

   A diesel engine run pump is installed for pumping saline water from well
   to feed the solar stills. A multi-blade vertical axis windmill has also
   been installed as an alternative system of pumping water. In collaboration
   with Central Electronics Ltd., Sahibabad, U.P. a solar cell operated pump
   of 120 watt capacity has also been installed in October, 1979. It has
   16 standard solar cell panels, two lead acid truck batteries of 12 V each
   connected in series and a D.C. centrifugal pump. This pump is now used
   for pumping product water as well as saline water from well at a rate of
   30 l/min. Thus for most of the period in the year the plant operates
   with natural sources of energy, namely wind and solar.

6. Monthly performance:

   The daily production from the plant for the year 1978-79 is shown in
   Table A-2. With the passage of time decrease in output has been observed.
   The reasons are listed below:

   (i)   Non-operation of some units due to leakages in bottom.
   (ii) Vapor leakages through leaking joints and broken glass sheets.
         The sealant used needs regular checking and resealing of joints.
   (iii) Exhorbitant growth of algae in some stills.
   (iv) Overflowing of distillate channels in some stills.

   The last problem is created due to accumulation of dust and other
   material in the channel finding its way through glass joints and the gap
   between lower edge and the channel. Dust nuisance is a common problem in
   India and better sealants are needed to avoid blockage of channels.

   Data collected during May, 1981 has shown that some individual stills
   have productivity above 4 litres/m^2 day.

124 Solar Distillation

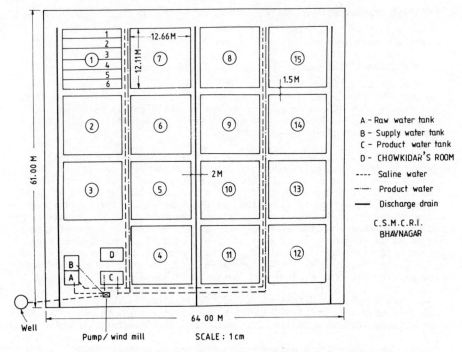

Fig. A-1. Layout plan of Awania solar still complex.

Fig. A-2. Awania solar still plant: pictorial view.

Appendix A

TABLE A-2. Monthly Performance of Solar Stills Plant at Awania
(Area of 90 stills - 1866.6 sq.m)

Month	Climate data for Bhavnagar			Average production liters/day actual	Water consumed liters/day
	Radiation intensity cal/cm² x day	Ambient temperature °C	Wind velocity KMPH		
1978					
March	582	26.0	9.1	6175	3260
April	615	30.9	9.4	6515	4260
May	657	32.7	11.4	6760	3360
June	486	30.8	11.5	4920	3900
July	377	29.1	10.5	2880	2730
August	326	28.3	11.6	3050	2390
September	491	28.6	8.7	3030	2700
October	484	27.3	6.5	3920	3730
November	405	26.3	7.3	4630	4530
December	406	21.9	6.5	3900	4010
1979					
January	403	21.2	7.4	3490	3990
February	510	21.9	7.3	3750	4325
Annual average	479	27.1	8.9	4420	3599

7. Reaction of the villagers about the solar still:

   Villagers use the water produced in this plant. But for the present population which is more than 1600 souls the water produced is insufficient and ladies living away from the plant still prefer to fetch water from well which is in the center of village. This preference can be overcome only by supplying product water from the plant through a tank constructed in the center of village.

8. Economics:

   When the plant was constructed in 1977-78 the investment cost inclusive of piping, pumps, tanks etc. was around Rs. 130-135 per sq.m. The solar stills are capital intensive as compared to other desalination plants. But they are simple to operate and maintain. The problems of maintenance are linked with size of plant and hence installation of solar stills of still larger capacities even under ideal conditions may not be very attractive both from investment and maintenance point of view.

9. Scientists involved in this project:

   Professor K. S. Rao, Dr. S. D. Gomkale, Shri G. L. Natu and Shri H. D. Goghari.

10. Miscellaneous:

    The design adopted in the construction needs to be modified to overcome problems faced in operation and maintenance. A still with modified construction has been constructed at Awania and it is giving better performance. However, with increasing costs of materials the investment costs are now around Rs. 300 per sq.m. and this factor definitely weighs against solar stills. There is a need to develop design for stills with higher output using readily available materials, better sealants etc. Though solar stills can desalt saline water with salinity higher than 10,000 ppm (a limit for desalination processes based on membranes) the availability of saline water of uniform quality and in adequate quantity needs to be checked prior to installation of plants.

APPENDIX B

CONVECTIVE HEAT TRANSFER FROM BASIN LINER TO WATER MASS

The convective heat transfer coefficient for heat flow from the horizontal basin liner (hottest region in the still, no dye is used) to the water mass in the basin, and vice-versa (dye is used in basin water) is determined from the following relation (McAdams, 1954)

$$Nu = C(Gr.Pr)^n$$

where the values of C and n for different cases are as follows:

C	n	direction of heat flow
0.54	1/4	upward (hot surface facing upward)
0.27	1/4	downward (hot surface facing downward)

After substituting the values of Nu, Gr and Pr, the convective heat transfer coefficient between basin liner and water can be written as

$$h_3 = \frac{C\,k_f}{x_1} \left[\frac{x_1^3 \rho_f^2 g \beta \Delta T}{\mu_f^2} \cdot \frac{C_{pf}\mu_f}{k_f} \right]^{\frac{1}{4}}$$

where the units of h_3 are W/m^2 $^\circ$C.

11. NOMENCLATURE

A	Area, m^2
b	Breadth of solar still, m
C	Specific heat, J/Kg $^{\circ}$C
C_{pa}	Specific heat of air at constant pressure, J/Kg $^{\circ}$C
C_{pf}	Specific heat of saturated air at constant pressure, J/Kg $^{\circ}$C
$C_{p\omega}$	Specific heat of water at constant pressure, J/Kg $^{\circ}$C
C_1, C_2	Specific heat of insulation and ground or concrete respectively, J/Kg $^{\circ}$C
d	Depth of water in the still, m
G_r	Grashof number
g	Acceleration due to gravity, m/sec
H_s	Incident solar radiation on glass cover per unit area per unit time, W/m^2
h	Heat transfer coefficient from water surface to glass cover, W/m^2 $^{\circ}$C
h_{ca}	Convective heat transfer coefficient from glass cover to ambient W/m^2 $^{\circ}$C
h_b	Overall heat transfer coefficient from water to atmosphere through bottom and sides of stills, W/m^2 $^{\circ}$C
h_D	Mass transfer coefficient $Kg/(hr\ m^2)/(Kg/m^3)$
h_{eff}	Evaporative heat transfer coefficient from water to glass cover, W/m^2 $^{\circ}$C
h_{fc}	Enthalpy of liquid water at cover temperature, J/Kg
$h_{g\omega}$	Enthalpy of water vapor at basin temperature, J/Kg
h_{ra}	Radiative heat transfer coefficient from glass cover to ambient, W/m^2 $^{\circ}$C
h_3	Convective heat transfer coefficient from basin liner to water, W/m^2 $^{\circ}$C

h_4	Convective heat transfer coefficient from bottom of the still to atmosphere, W/m² °C
h_5	Convective heat transfer coefficient from lower glass cover to upper basin water, W/m² °C
h_6	Combined convective, evaporative and radiative heat transfer coefficient from lower basin water to lower glass cover, W/m² °C
K	Thermal conductivity, W/m °C
K_g	Thermal conductivity of glass, W/m °C
K_1, K_2	Thermal conductivity of insulation and ground or concrete respectively, W/m °C
k_f	Thermal conductivity of saturated air, W/m °C
L	Thickness of insulation, m
L'	Length of still in y-direction, m
$L_{s\omega}$	Least width dimension of the solar still, m
$\mathcal{L}$	Latent heat of vaporisation of water, J/Kg
ℓ_g	Thickness of glass, m
ℓ_ω	Depth of water in basin, m
M	Molecular weight, Kg
$\dot{M}_a$	Mass of air transferred per unit area per unit time by free convection, Kg/m² hr
M_e	Daily productivity of distilled water, Kg/m² day
M_g	Heat capacity of glass per unit area, J/m² °C
M_{gL}	Heat capacity of lower glass in double basin still, J/m² °C
M_{gu}	Heat capacity of upper glass in double basin still, J/m² °C
M_R	Resultant water mass, i.e. $(M_{\omega o} - \dot{m}_e)$, Kg/m²
$M_{\omega L}$	Heat capacity of lower basin water in a double basin still, J/m² °C
$M_{\omega o}$	Heat capacity of water mass in basin, J/m² °C
$M_{\omega u}$	Heat capacity of upper basin water, J/m² °C
$\dot{m}_e$	Instantaneous distillation rate, Kg/hr m² (mass evaporated per unit area per unit time)
m_ω	Cumulative distillate product in time t, Kg/m²
$\dot{m}_\omega$	Flow rate of water, Kg/sec
n	An integer
N_u	Nusselt number
P	Partial vapor pressure at temperature T, pa
P_a	Partial pressure of water vapor at atmospheric temperature, pa
P_g	Partial pressure of water vapor at glass temperature, pa
P_r	Prandtl number
P_T	Total pressure of a mixture of air and water vapor in a closed container in equilibrium, pa
P_ω	Partial pressure of water vapor at water temperature, pa

Nomenclature

Q_e	Total amount of solar energy used for evaporation, J/m^2 day
Q_t	The total amount of solar radiation incident on the still cover, J/m^2 day
$\dot{Q}(t)$	Amount of heat flux per unit area coming into the room through roof, W/m^2
q	Amount of heat transferred per unit area per unit time from water to glass, W/m^2
q_a	Total heat transferred per unit area per unit time from glass to ambient, W/m^2
q_{ca}	Heat transferred from glass cover to atmosphere by convection, W/m^2
q_{ins}	Heat transferred from basin liner to atmosphere by conduction through bottom insulation, W/m^2
q_L	Total heat transferred from lower basin water to middle glass, W/m^2
q_u	Total heat transferred from upper basin water to upper glass, W/m^2
q_{ra}	Heat transferred from glass cover to atmosphere by radiation, W/m^2
q_ω	Heat transferred from absorbing surface to water, W/m^2
q'	Energy needed to heat saline water, J/m^2
R	Universal gas constant
R_g	Reflection coefficient of glass cover
S_s	Shape factor for calculating heat loss to the ground, dimensionless
T	Temperature, $°C$
T'	Temperature, $°K$
T_F	Final water temperature, $°C$
T_i	Water temperature prior to being heated, $°C$
T_{sky}	Sky temperature, $°C$
T_ω	Water temperature, $°C$
$T_{\omega o}$	Initial water temperature, $°C$
t	Time, hr
U_L	Overall heat transfer coefficient from absorbing surface of the still to ambient, $W/m^2 \ °C$
V	Volume, m^3
ϑ	Wind speed, m/sec
ϑ_f	Fluid velocity, m/sec
W	Brine flow rate, Kg/m^2 hr
x_1	Spacing between water surface and glass cover, m
x	Position co-ordinate vertically downward, m
α_g	Absorption coefficient of glass
α_L	Absorption coefficient of basin liner
α_ω	Absorption coefficient of water
β'	Coefficient of volumetric thermal expansion, $°C^{-1}$

ϵ_1, ϵ_2	Emissivity of inside surfaces of two infinite parallel planes (in still $\epsilon_1 = \epsilon_g$ and $\epsilon_2 = \epsilon_\omega$)
$\theta(x,t)$	Temperature distribution, $^\circ C$
θ_R	Constant room air temperature, $^\circ C$
$\theta_1\|_{x=0} = \theta_{basin}$	
$\epsilon_g, \epsilon_\omega$	Emissivity of glass and water respectively
η	Efficiency of still
μ	Coefficient of viscosity, Kg/m sec
ρ	Partial mass density of water vapor, Kg/m^3
ρ_ω	Density of water, Kg/m^3
ρ_o	Density, Kg/m^3
ρ_f	Density of saturated air, Kg/m^3
ρ_1, ρ_2	Density of insulation and ground or concrete respectively Kg/m^3
σ	Stefan-Boltzmann constant, 5.6697×10^{-8} W/m^2 $^\circ K^4$

$\tau_1 = (1-R_g)\alpha_g$
$\tau_2 = (1-R_g)(1-\alpha_g)\alpha_\omega$ For single basin solar still
$\tau_3 = (1-R_g)(1-\alpha_g)(1-\alpha_\omega)\alpha_b$

$\tau_1 = (1-R_g)\alpha_g$
$\tau_2 = (1-R_g)(1-\alpha_g)\alpha_\omega$
$\tau_3 = (1-R_g)(1-\alpha_g)(1-\alpha_\omega)\alpha_g$ For double basin solar still
$\tau_4 = (1-R_g)(1-\alpha_g)^2(1-\alpha_\omega) \times \alpha_\omega$
$\tau_5 = (1-R_g)(1-\alpha_g)^2 \times (1-\alpha_\omega)^2 \alpha_b$

ω	2π/period, Sec^{-1}

Subscripts

o, ∞	refer to conditions at the evaporation interface and far away respectively
cω, eω, rω	refer to by convection, evaporation and radiation from water surface to glass cover, respectively
eL, eu	refer to evaporation from lower and upper basin respectively
f	refers to saturated air
g	refers to property of glass cover
g$_i$, g$_o$	refer to inside and outside glass surfaces
in	refers to inlet water temperature
u, L	refer to upper and lower basin of a double basin solar still
ω, a	refer to the diffusing (water vapor) and inert (air/gases respectively
1, 2	refer to inside surface of two infinite parallel planes

12. REFERENCES

Abbot, C.G. (1930) *Smithsonian Inst. Misc. Coll.*, 98(5), Pub. No. 3530.
Abbot, C.G. (1944) *Smithsonian Inst. Ser.*, 2; U.S. Patent No. 2, 141, 330 (Dec. 27, 1938).
Achilov, B.M., Zhuraev, T.D. and Akhtamov, R.A. (1972) Results of year long tests of tilted step solar stills, *Gelitekhnika*, 8, 78.
Achilov, B.M., Zhuraev, T.D. and Akhtamov, R. (1973a) Choice of material and technology for solar still, *Geliotekhnika*, 9 (5), 39.
Achilov, B.M., Zhuraev, T.D. and Akhtamov, R. (1973b) Test on a portable solar still, *Geliotekhnika*, 9 (6), 51.
Achilov, B.M., Akhtamov, R.A., Zhuraev, T.D. and Ten, M.Kh. (1976) A regenerative tray type solar still, *Geliotekhnika*, 12 (2), 22.
Ahmadzadeh, J. (1978) Solar earth water stills, *Solar Energy*, 20 (5), 387.
Akhtamov, R.A., Achilov, B.M., Kamilov, O.S. and Kakharov, S. (1978) Study of regenerative inclined-stepped solar still, *Geliotekhnika*, 14 (4), 51.
Akinsete, V.A. and Duru, C.U. (1979) A cheap method of improving the performance of roof type solar still, *Solar Energy*, 23 (3), 271.
ASHRAE Handbook of Fundamentals (1967) American Society of Heating, Refrigerating and Air-conditioning Engineers, New York, p.42.
Baibutaev, K.B., Achilov, B.M. and Kamaeva, G. (1970) Effect of salt concentration on the evaporation process in solar stills, *Geliotekhnika*, 6 (2), 83.
Bairamov, R., Ushakova, A.D. and Agadzhanov, V. (1979) Supplying water to desert pastures by means of solar desalination plants, *Geliotekhnika*, 15 (5), 78.
Bartali *et al.* (1976) Chimney and heated head solar still, *Heliotechnique and Development*, II, 431.
Baum, V.A. and Bairamov, R. (1964) Heat and mass transfer processes in solar stills of hot-box type, *Solar Energy*, 8, 78.
Baum, V.A. and Bairamov, R. (1966) Prospects of solar stills in Turkmenia, *Solar Energy* (United States of America), No. 1.
Baum, V.A., Bairamov, R. and Malevski, Yu. (1967) Possibilities of using solar energy in reclaiming deserts, *Problems of Desert Reclamation*, Report No. 5, Ashkhabad.
Baum, V.A. *et al.* (1968) In *Proc. VIIth World Power Conference* (Moscow).
Baum, V.A., Bayaramov, R.B. and Malevsky, Y.M. (1970) The solar still in the desert, *Proc. International Solar Energy Congress*, Melbourne, p.426.

Bloemer, J.W., Allen, J.M. and Eibling, J.A. (1961a) *Twelfth Quarterly Progress Report of Solar Sea-water Stills*, Office of Saline Water, Battelle Memorial Institute, Columbus, Ohio.

Bloemer, J.W., Collin, R.A. and Eibling, J.A. (1961b) *Study and Field Evaluation of Solar Sea-water Still*, Report No. 50, Office of Saline Water, Battelle Memorial Institute, Columbus, Ohio.

Bloemer, J.W., Irwin, J.R. and Eibling, J.A. (1964) *Second Two-years Progress on Study and Field Evaluation of Solar Sea-water Stills*, Office of Saline Water, Battelle Memorial Institute, Columbus, Ohio.

Bloemer, J.W., Irwin, J.R. and Lof, G.O.G. (1965a) Paper presented at Solar Energy Society Annual Meeting, March 1965.

Bloemer, J.W., Irwin, J.R., Eibling, J.A. and Lof, G.O.G. (1965b) A practical basin type solar still, *Solar Energy*, 9, 197.

Bowen, I.S. (1926) The ratio of heat loss by conduction and by evaporation for any water surface, *The Physical Review*, 27 (2nd series), 779.

Coffey, J.P. (1975) Vertical solar distillation, *Solar Energy*, 17, 373.

Cooper, P.I. (1969a) Digital simulation of transient solar still processes, *Solar Energy*, 12, 313.

Cooper, P.I. (1969b) The absorption of radiation in solar stills, *Solar Energy*, 12, 333.

Cooper, P.I. (1970) The transient analysis of glass covered solar still, Ph.D. Thesis, University of Western Australia, Australia.

Cooper, P.I. (1972) Some factors affecting the absorption of solar radiation in solar stills, *Solar Energy*, 13, 373.

Cooper, P.I. (1973a) Digital simulation of experimental solar still data, *Solar Energy*, 14, 451.

Cooper, P.I. (1973b) Maximum efficiency of a single effect solar still, *Solar Energy*, 15, 205.

Cooper, P.I. and Appleyard, (1967) The construction and performance of a three effect, wick type, tilted solar still, *Sun at Work*, 12 (1), 4.

Cooper, P.I. and Read, W.R.W. (1974) Design philosophy and operating experience for Australian solar stills, *Solar Energy*, 16, 1.

Daniels, F. (1965) Construction and tests of small solar stills, Solar Energy Laboratory, University of Wisconsin, *Proc. Solar Energy Society Annual Meeting*, Phoenix, Arizona, p.1.

Datta, R.L., Gomkale, S.D., Ahmed, S.Y. and Datar, D.S. (1965) Evaporation of sea water in solar stills and its development for desalination, *Proc. First Int. Symposium on Water Desalination*, Washington, D.C. (Oct.).

Delyannis, A.A. (1965) Solar stills provide an island's inhabitants with water, *Sun at Work*, 10 (1), 6.

Delyannis, A. and Delyanis, E. (1973) Solar distillation plant of high capacity, *Proc. 4th Int. Symp. on Fresh Water from Sea*, 4, 487.

Donald, Q.K. (1950) *Process Heat Transfer*, Kogalusha Company Limited, Tokyo, McGraw Hill Book Company Inc., New York, p.384.

Duffie, J.A. and Beckman, W.A. (1974) *Solar Energy Thermal Processes*, John Wiley and Sons, New York.

Dunkle, R.V. (1961) Solar water distillation; the roof type still and a multiple effect diffusion still, *International Developments in Heat Transfer*, A.S.M.E., Proč. *International Heat Transfer*, Part V, University of Colorado, p.895.

Edlin, F.E. (1965) Air supported solar still, E.I. du Pont de Nemours, United States Patent No. 3,174,915 Filed July 1962, Serial No. 211, 942.

Falvey, H.T. and Todd, C.J. (1980) Concentric tube solar still, private communication, U.S. Bureau of Reclamation Engineering and Research Centre, P.O. Box 25007, Denver Federal Center, Denver, Colorado 80225.

Frick, B. (1970) Some new considerations about solar stills, *Proc. International Solar Energy Congress*, Melbourne, p.395.

References

Frick, G. and Sommerfeld, J.V. (1973) Solar stills of inclined evaporating cloth, *Solar Energy*, 14, 427.

Garg, S.K., Gomkale, S.D., and Datta, R.L. (1966) Use of solar energy for production and supply of water from salt water, Presented at Symposium on Community Supply and Waste Disposal, at CPHERI, India (Dec. 19-20).

Garg, H.P. and Mann, H.S. (1976) Effect of climatic, operational and design parameters on the year round performance of single sloped and double sloped solar still under Indian and arid zone conditions, *Solar Energy*, 18, 159.

Ginnings, D.C. (1948) Multiple effect solar still, U.S. Patent No. 2, 445.

Gomkale, S.D. (1969,80) Solar distillation, CSMCRI, Bhavnagar, India (private communication).

Harding, J. (1883) Apparatus for solar distillation, *Proc. Institute of Civil Engineers*, 73, 284.

Hay, H.R. (1966) V-cover solar stills, *Sun at Work*, II (2nd quarter) (2), 6.

Henrik, W. (1972) Fresh water from sea water, distillation by solar energy, *Solar Energy*, 13 (4), 439.

Hirschmann, J.R. and Roefler, S.K. (1970) Thermal inertia of solar stills and its influence on performance, *Proc. International Solar Energy Congress*, Melbourne, p.402.

Hodges, C.N., Thompson, T.L., Groh, J.E. and Frieling, D.H. (1966b) *Solar Distillation Utilizing Multiple Effect Humidification*, Office of Saline Water, U.S. Department of the Interior Research and Development, Report No. 194.

Howe, E.D. (1961) Solar distillation research at the University of California, U.N. Conference on New Sources of Energy, Session III E, E/Conf. 35/5/29, Rome, p.1.

Howe, E.D. (1964) Solar distillation problems in developing countries, paper 64-WA/SOL-7, ASME Meeting, New York.

Howe, E.D. and Tleimat, B.W. (1974) Twenty years of work on solar distillation at the University of California, *Solar Energy*, 16, 97.

Jacob, M. (1949) *Heat Transfer*, Wiley & Sons, New York, Vol. I.

Jacob, M. (1957) *Heat Transfer*, Wiley & Sons, New York, Vol. II.

Kausch, O. (1920) *Die unmittelbare ausnutzung der Sonnenenergia*, Weimar, Carl Steinert.

Keller, J.D. (1928) The flow of heat through furnace hearths, *Trans. ASME*, Pittsburgh, Pa. p.111.

Kettani, M.A. (1979) Review of solar desalination, *Sun World*, 3 (3), 76.

Khatry, A.K., Sodha, M.S. and Malik, M.A.S. (1978) Periodic variation of ground temperature with depth, *Solar Energy*, 20, 425.

Kobayashi, M. (1963) A method of obtaining water in arid land, *Solar Energy*, 7 (3), 93.

Lavoisier, A.L. (1862) *Oeuvres de Lavoisier*, Son Excellence le ministre de l'instruction publique et de cultes, 3, Takle 9.

Lawand, T.A. (1968) *Engineering and Economic Evaluation of Solar Distillation for Small Communities*, Tech. Report No. MT-6, Brace Research Institute of McGill University, Canada.

Lewis, W.K. (1922) The evaporation of a liquid into a gas, *ASME Trans.* 44, 325.

Lewis, W.K. (1933) Evaporation of a liquid into a gas-a correction, *Mechanical Engineering*, 55, 567.

Lobo, P.C. and Araujo, S.R.D. (1977) Design a simple multi-effect basin type solar still, *Proc. International Solar Energy Congress*, New Delhi, p.2026.

Lof, G.O.G. (1961a) Fundamental problems in solar distillation, *Solar Energy*, 5, 35.

Lof, G.O.G., Eibling, J.A. and Bloemer, J.W. (1961b) Energy balances in solar distillation, *Am. Inst. Chem. Engrs.*, 7, 641.

Lof, G.O.G. (1962) Final Report on consultant contract of Feb. 1961, including Progress Report for Jan-March, 1962 (unpublished report to Office of Saline Water), Washington.

Lof, G.O.G. (1966) Solar desalination, Chapter 5 in *Principles of Desalination*, edited by K.S. Spiegler, Academic Press, New York.
MacLeod, L.H. and McCracken (1961) Univ. California Rept. 27 (series 75), Berkeley.
Malik, M.A.S., Puri, V.M. and Aburshaid, H. (1978) Use double stage solar still for nocturnal production, *Proc. 6th International Symposium Fresh Water from the Sea*, 2, 367.
Malik, M.A.S. and Tran, V.V. (1973) A simplified mathematical model for predicting the nocturnal output of a solar still, *Solar Energy*, 14, 371.
McAdams, W.C. (1954) *Heat Transmission*, 3rd edition, McGraw Hill, New York.
McCracken, H.W. (1965) Solar still pans; the search for inexpensive and durable materials, Paper presented at Solar Energy Society Annual Meeting.
Menguy, G., Benoit, M., Louat, R., Makki, A. and Schwartz (1980) New solar still design and experimentation (The wiping spherical still), private communication, Group d'Etudes Thermiques et Solaires, Universite' Claude Bernard 43 Bd du 11 Novembre 1918, 69622 Villeurbanne - Cedex France.
Morse, R.N. and Read, W.R.W. (1968) A rational basis for the engineering development of a solar still, *Solar Energy*, 12, 5.
Morse, R.N., Read, W.R.W. and Trayford, R.S. (1970) Operating experiences with solar stills for water supply in Australia, *Solar Energy*, 13, 99.
Mouchot, A. (1869) *La Chaleur Solavie et ses Applications Industrielles*, Gauthier - Villars, Paris, p.1.
Moustafa, S.M.A. and Brusewitz (1979) Direct use of solar energy for water desalination, *Solar Energy*, 22, 141.
Nayak, J.K., Tiwari, G.N. and Sodha, M.S. (1980) Periodic theory of solar still, *Int.J. of Energy Research*, 4, 41.
Nebbia, G. and Menozzi, G. (1966) *A Short History of Water Desalination*, Acqua Dolee Dal Mare, II Inchiesta Internazionole Milano, p.129.
Norov, E.Zh., Achilov, B.M. and Zhuraev, T.D. (1975) Results of tests on solar film-covered stills, *Geliotekhnika*, 11 (3/4), 130.
Office of Saline Water (1966) *Final Three Years Progress Report on Study and Field Evaluation of Solar Sea-water Still*, Battelle Memorial Institute, Colombus, Ohio, Report No. 190.
Oltra, F. (1972) *Saline Water Conversion and Its Stage of Development in Spain*, Publication of J.E.N., Madrid.
Oztoker, U. and Selcuk, M.K. (1971) *Theoretical Analysis of a System Coupling a Solar Still with a Controlled Environment Greenhouse*, ASME paper No. 71/WA/SOL/9, Washington D.C. 28.
Pasteur, F. (1928) (title unknown) *Compt. Rend.*, 30, 187.
Qasim, S.R. (1978) Treatment of domestic sewage by using solar distillation and plant culture, *Journal of Environmental Science and Health*, 13 (8), 615.
Rajvanshi, A.K. and Hsieh, C.K. (1979) Effect of dye on solar distillation: Analysis and experimental evaluation, *Proc. Int. Congress of ISES*, Georgia, P-20,327.
Sayigh, A.A.M. (ed.) (1977) *Solar Energy Engineering*, Academic Press.
Schmidt, Ernst (1969) *Properties of Water and Steam in SI Units*, Springer-Verlag, Berlin.
Selcuk, M.K. (1970) The effect of solar radiation on the energy balance of a controlled environment greenhouse, ASME Annual Winter Meeting, New York, Paper No. 70/WA/5063.
Selcuk, M.K. (1971) Analysis, design and performance evaluation of controlled environment greenhouse, *Trans ASHRAE*, No. 2172.
Sharpley, B.F. and Boelter, L.M.K. (1938) Evaporation of Water into quiet air, *Indus.Eng.Chem.* 30 (10), 1125.
Sodha, M.S., Kumar, A., Singh, Usha and Tiwari, G.N. (1980a) Transient analysis of solar still, *Energy Conversion*, 20 (3) 191.

Sodha, M.S., Kumar, A., Tiwari, G.N. and Tyagi, R.C. (1981a) Simple multiple-wick solar still: Analysis and performance, *Solar Energy*, 26 (2), 127.

Sodha, M.S., Kumar, A., Srivastava, A. and Tiwari, G.N. (1980b) Thermal performance of still on roof system, *Energy Conversion*, 20 (3), 181.

Sodha, M.S., Nayak, J.K., Tiwari, G.N. and Kumar, A. (1980c) Double basin solar still, *Energy Conversion*, 20 (1), 23.

Sodha, M.S., Kumar, A., Tiwari, G.N. and Pandey, G.C. (1980d) Effect of dye on thermal performance of solar still, *Applied Energy*, 7, 147

Sodha, M.S., Kumar, A., Singh, Usha and Tiwari, G.N. (1980e) Further studies on double basin solar still, *Int.J. Energy Research* (In Press).

Sodha, M.S., Khatry, A.K. and Malik, M.A.S. (1978) Reduction of heat flux through a roof by water film, *Solar Energy*, 20, 189.

Sodha, M.S., Kumar, A. and Tiwari, G.N. (1981b) Utilization of waste hot water for distillation, *Desalination* (In Press).

Soliman, H.S. (1972) Effect of wind on solar distillation, *Solar Energy*, 13, 403.

Soliman, H.S. (1976) *Solar Still coupled with a Solar Water Heater*, Mosul University, Mosul, Iraq, p.43.

Talbert, S.G., Eibling, J.A. and Lof, G.O.G. (1970) *Manual on Solar Distillation of Saline Water*, R & D Progress Report No. 546, US Dept. of the Interior.

Telkes, M. (1945) *Solar Distiller for Life Rafts*, United States Office of Science, R & D Report No. 5225, P.B.21120.

Telkes, M. (1956) In *Proc. World Symposium on Applied Solar Energy*, Stanford Research Institute, Menlo Park, California, p.73.

Threlkeld, J.L. (1970) *Thermal Environmental Engineering*, Prentice-Hall Inc., New Jersey.

Tinaut, D., Echaniz, G. and Ramos, F. (1978) Materials for a solar still greenhouse, *Optica Pura Y Aplicada*, 11, 59.

Tleimat, B.W. and Howe, E.D. (1966) Nocturnal production of solar distiller, *Solar Energy*, 10 (2), 61.

Tleimat, B.W. and Howe, E.D. (1967) Comparison of plastic and glass condensing covers for solar distillers, *Proc. Solar Energy Society*, Annual Conference, Phoenix, Arizona, or *Solar Energy*, 12, 293 (1969).

Trombe, F. and Foex, M. (1961) Utilization of solar energy for simultaneous distillation of brackish water and air-conditioning of hot houses in arid region, U.N. Conference of New Source of Energy Paper 35/3/64 Revised Rome 11, 6.

Umarov, G. Ya, Asamov, M.K., Achilov, B.M., Sarros, T.K., Norov, E.Zh. and Tsagaraeva, N.A. (1976) Modified polyethylene films for solar stills, *Geliotekhnika*, 12 (2), 29.

United Nations (1964) U.N. Publ. 64, IIB. 5, United Nations New York.

United Nations (1970) *Solar Distillation*, Dept. of Economics and Social Affairs, New York.

Wong, H.Y. (1977) *Heat Transfer for Engineers*, Longmans, London.

13. SELECTED BIBLIOGRAPHY ON SOLAR DESALINATION

Abbas, M.A. et al., Experimental studies on basin-type solar distillers in Iraq, *Revue d' Heliotechnique*, 1, 35 (1977).
Abbot, C.G., *The Sun and the Welfare of Man*, Smithsonian Scientific Series (1934).
Achilov, B.M., Umarov, G. Ya. and Baibutayev, G.B., Economical efficiency of a solar still introduced into the Republic water supply system, *Applied Solar Energy (Geliotekhnika)*, 6(4), 87 (1970).
Achilov, B.M., Umarov, G. Ya., Baibutaev, K.B., Zakhidev, R.A. and Zhuraev, T.D., Investigation of an industrial-type solar still, *Applied Solar Energy (Geliotekhnika)*, 7(2), 45 (1971).
Achilov, B.M., Vardiashvilli, A., Umarov, G.Ya., Baibutayev, K.B., Khatamov, S. and Djuraev, T.D., Investigation of inclined and stepped solar stills, *Applied Solar Energy (Geliotekhnika)*, 7(5), 48 (1971).
Achilov, B.M., Vardiashvilli, A.B., Umarov, G.Ya, Baibutayev, C.B. and Djuraev, T.D., Comparison tests of industrial solar stills in the Kizil-Kum deserts in the Uzbek Republic, *Applied Solar Energy (Geliotekhnika)*, 7(5), 64 (1971).
Achilov, B.M., A study of solar stills of the trough-shaped and inclined stepped type and their possible application as sources of water in agriculture in Uzbekistan, Abstract of Dissertation, Tashkent (1971). In Russian.
Achilov, B.M., Umarov, G.Ya., Zhuraev, T.D. and Akhtamov, R., Use of solar stills for pasture water supply in Uzbekistan, *Proc. First All-Union Scientific-Technical Conference on Replenishable Sources of Energy*, No. 1, Energiya (1972). In Russian.
Achilov, B.M., Umarov, G.Ya., Baybutaev, K.B. and Djuraev, T., Annual operational data on industrial solar still, *Applied Solar Energy (Geliotekhnika)*, 8(1), 45 (1972).
Achilov, B.M., Jurayev, T.D. and Akhtamov, R.A., Result of a year's testing of inclined and stage type solar distilling units, *Applied Solar Energy (Geliotekhnika)*, 8(3), 59 (1972).
Achilov, B.M., Zhuraev, T.D. and Akhtamov, R., Choice of materials and technology for solar stills, *Applied Solar Energy (Geliotekhnika)*, 9(5), 28 (1973).
Achilov, B.M., Zhuraev, T.D. and Akhtamov, R., Tests on a portable solar still, *Applied Solar Energy (Geliotekhnika)*, 9(6), 92 (1973).
Achilov, B.M. and Zhuraev, T.D., Results of tests on solar film-covered stills, *Applied Solar Energy (Geliotekhnika)*, 10(4), 104 (1974).

Selected Bibliography

Achilov, B.M., Akhtamov, R.A., Zhuraev, T.D. and Ten, M.Kh., A regenerative tray-type solar still, *Applied Solar Energy (Geliotekhnika)*, 12(2), 16 (1976).

Achilov, B.M., Norov, E.Zh. and Kamaeva, G.B., Chemical investigation of water distilled by film solar plants, *Applied Solar Energy (Geliotekhnika)*, 12(4), 68 (1976).

Achilov, B.M., Kamilov, O.S., Odinaev, A. and Zhuraev, T.D., Test results of sections of an industrial solar distiller, *Applied Solar Energy (Geliotekhnika)*, 15(2), 78 (1979).

Achilov, B.M., Bobrovnikov, G.N., Kakharov, S., Zhuraev, T.O. and Kamilov, O.S., Determination of basic structural parameters of solar stills of inclined-stage type, *Applied Solar Energy*, 15(5), 80 (1979).

Achilov, B.M., Kakharov, S., Zhuraev, T.O. and Norov, E., Test results for a combination distilling plant, *Applied Solar Energy*, 16(2), 46 (1980).

Ahmed, S.Y., Gomkale, S.D., Datta, R.L. and Datar, D.S., Scope and development of solar stills for water desalination in India, *Desalination*, 5, 64 (1968).

Ahmed, Syed, Faruq, Nocturnal and daytime performance of a single stage solar still, M.Sc. Thesis, Colorado State University, Fort Collins, Colorado (1979).

Akhtamov, R.A., Achilov, B.M., Kamilov, O.S. and Kakharov, S., Study of regenerative inclined stepped solar still, *Applied Solar Energy (Geliotekhnika)*, 14(4), 51 (1978).

Akhtamov, R.A., Achilov, B.M., Kakharov, S., and Zhuraev, T.D., Annual operating data for type SOV-1000 stepped oblique type industrial solar stills, *Applied Solar Energy*, 14(2), 47 (1978).

Akinsete, V.A. and Duru, C.U., A cheap method of improving the performance of roof type solar still, *Solar Energy*, 23(3), 271 (1979).

Alward, R. and Lawand, T.A., *The Installation Costs of the Auxiliaries for a Small Desalination Unit in an Isolated Community*, Brace Research Institute, McGill University, Technical Report No. T-46, 12 pages (April 1968).

Alward, R. and Lawand, T.A., *The Cost of Supplying Sea Water for Sanitary Purposes in an Isolated Community*, Brace Research Institute, McGill University, Technical Report No. T-40, 9 pages (May 1968).

Alward, R., Acheson, A. and Lawand, T.A., The integration of solar stills into minimum cost dwellings for arid areas, *Proc. International Congress, 'The Sun in the Service of Mankind'*, UNESCO, Paris (1973).

Ambroggi, R., Investigations conducted in Morocco about the demineralization of saline and brackish waters through solar distillation, Colloque UNESCO, Iran, Tehran (1958).

Anand, S.P., A modified basin-type solar still, *Defence Scientific Journal*, 28, 5 (1978).

Anand, S.P., Use of black in free form for absorption of solar heat, *Proc. ISES, Belgrade*, 20, 1 (1978).

Anand, S.P., Concerted efforts to improve the efficiency of multi-surface solar still, *Proc. National Solar Energy Convention*, I.I.T. Bombay, Paper 45, 238 (1979).

Annaev, A., Bairamov, R. and Rybakova, L.E., The effect of wind speed and direction on the output of a solar still, *Applied Solar Energy (Geliotekhnika)*, 7(4), 74 (1971).

Anon., Sterilization of water by solar heat (In French), *Administr. Locale*, 54, 1059-1060 (April-June 1930).

Anon., Still converts salt water into safe drinking water, *Sci. News Letters*, 47(6), 88 (Feb. 10, 1945).

Anon. Solar energy to provide water supply for island, *Heat. and Vent.*, 45(8), 99 (Aug. 1948).

Anon., Solar stills, *Modern Plastics*, 28, 68-69 (June 1950).

Anon., Low-cost solar distillation predicted, *Chem. Eng. News*, 29(16), 1492 (Apr. 16, 1951).

Anon., Improved solar still, *Int. Chem. Eng.*, 32(7), 314 (July 1951).
Anon., Sea water to drink, *Mod. Plastics*, 29(3), 90-91 (Nov. 1951).
Anon., Plastic solar stills, *Ind. & Eng. Chem.* 47(7), 9A-11A (July 1955).
Anon., We will drink the sea, *Sci.News Letter*, 68(6), 90-91 (Aug. 6, 1955).
Anon., Solar still operated at Mildura, *The Sun at Work*, 1(3), 1 (Sept. 1956).
Anon., Phoenix inventor designs solar still, *The Sun at Work*, 1(3), 14 (Sept. 1956).
Anon., Desalting the solar way, *Chem. Eng. News*, 6053-6054 (Dec. 10, 1956).
Anon., Solar distillation of saline water, *Science*, 124, 1287 (Dec. 28, 1956).
Anon., *A Standardized Procedure for Estimating Costs of Saline Water Conversion*, Office of Saline Water, Washington (1956).
Anon., Solar stills, *SPC Quart. Bull.*, 5 pages (Jan. 1957).
Anon., Plastics in solar stills - an interim report, *Mod. Plastics*, 34(10), 266-267 (June 1957).
Anon., Brewing a drink of sea water is easy - but still costly, *Bus. Week* (1480), 132-136 (Jan. 11, 1958).
Anon., Water desalting closer to commercial, *Chem. Week*, 31-32 (Aug. 30, 1958).
Anon., Solar test stills in operation, *The Sun at Work*, 4(2), 1, 12 (June, 1959).
Anon., Salt water conversion, *The Sun at Work*, 19, 4-7 (Second Quarter, 1960). (Reprinted portion of OSW Saline Water Conversion Report for 1959, *Review of Solar Distillation Processes*.
Anon., *Sea Water Conversion Program*, Berkeley Progress Report for the Year Ending June 30, 1960, Sea Water Conversion Laboratory, University of California, Berkeley, California, Series 75, Issue 23 (Aug. 1960).
Anon., Progress in saline-water conversion, *J. Amer. Water Works Assoc.*, 53(9), 1091-1105 (Sept. 1961).
Anon., Solar science project, *Sun at Work*, 13 (Second Quarter, 1963).
Anon., Progress in solar salt-water conversion, *Sun at Work*, 8-12 (Fourth Quarter, 1963). (Abst. from *Saline Water Conversion Report, 1962*, 1962 by OSW).
Anon., *Design of a Basin-type Solar Still*, U.S. Dept. of the Interior, Office of Saline Water, R&D Report, p. 112 (1964).
Anon., Sophisticated heat transfer upgrades solar water-desalting, *Chem. Eng.*, 71, 79-80 (June 22, 1964).
Anon., Emergency solar still desalts sea water, *Safety Maint.*, 130, 24 (Oct. 1965).
Anon. (JRM), Solar desalting gains acceptance in Greece, *Chemical Engineering*, 72(26), 42-43 (Dec. 20, 1965).
Anon., *Second Two Years Progress on Study and Field Evaluation of Solar Sea-Water Stills*, O.S.W. Research and Development Progress, Report No. 197 (1965).
Anon., *How to Make a Solar Still (plastic covered)*, Brace Research Institute, McGill University, Montreal (1965).
Anon., *Final Three Years Progress on Study and Field Evaluation of Solar Sea-water Stills*, O.S.W. Research and Development Progress, Report No. 190 (1966).
Anon., *A Study of Water Desalination Methods and Their Relevance to Australia*, Prepared for the Advisory Panel on Desalination, Australian Water Resources Council, Hydrological Series, No. 1 (1966).
Anon., Emergency solar still desalts seawater, *Sun at Work*, 9 (First Quarter, 1966). (From NASA Tech. Brief 65-1021 D, Manned Spacecraft Center, Houston, Texas 77001).
Anon., Greeks start up big solar distillation plant on Patmos, *Chemical and Engineering News*, 45(22), 22 (May 22, 1967).
Anon., Water, water, everywhere - solar distillation of water in South Pacific islands, *Medical Journal of Australia*, 551(18), 771 (1968).

Anon., Solar-stills producer sights big potential, newspaper article on Horace McCracken, The San Diego Union (July 1968).
Anon., Trends in desalination, *Nuclear Engineering*, 14(149), 854-858 (Oct. 1968).
Anon., An aqua-chem vapor compression sea water desalting unit is being sent to the Greek Island of Symi, *Desalting Digest*, 6 (July 1969).
Abbot, C.G. (Title unknown), *Smithsonian Inst. Misc. Coll.*, Publ. No. 3530, 98(5), (1930).
Anon., *Solar Distillation*, Department of Economics and Social Affairs, United Nations, New York, E.70 II B.I. (1970).
Anon., *Solar Distillation as a Means of Meeting Small-scale Water Demand*, United Nations, Department of Economics and Social Affairs, Publication No. 70 II B.1 (1970).
Anon., *Solar Energy in Developing Countries — Perspective and Prospect*, National Academy of Sciences, NTIS, PB-208-550 (1972).
Anon., Solar ponds as a heat source for low temperature, multieffect distillation plants, *Desalination*, 17, 289-302 (1975).
Anon., Solar desalination of saline water, *Huaxue Tongbao*, No. 6 p. 10, 18 (23 Nov., 1976). In Chinese.
Anon., Desalination by solar energy, *Proc. Int. Conf. on Energy Sources and Development*, Barcelona (October, 1977).
Anon., *The Design of Water Supply Systems based on Desalination*, United Nations Publication (Sales No. 68.II.B.20).
Apel'tsyn, I.E. and Klyachko, V.A., *Distillation of Water*, Moscow (1968). In Russian.
Appleyard, J.A., Solar energy research in Western Australia, *The Sun at Work*, 4-6 (Fourth Quarter, 1965).
Arce, Isaac R., *Historical Reports from Antofagasta* (In Spanish), 250, Antofagasta (1930).
Asamov, M.K., Achilov, B.M., Sarros, T.K., Norov, E.Zh. and Tsagaraeva, N.A., Modified polyethylene films for solar stills, *Applied Solar Energy (Geliotekhnika)*, 12(2), 23 (1976).
Association of Applied Solar Energy, *Bibliography of Solar Energy*, Second Edition (1959).
Avezov, R.R. and Norov, E.Zh., Thermal calculation procedures for solar distillers, *Applied Solar Energy (Geliotekhnika)*, 10(5), 20 (1974).
Avezov, R.R. and Norov, E.Zh., Theoretical study of the influence of wind velocity and direction and still non-hermiticity on its effectiveness, *Applied Solar Energy (Geliotekhnika)*, 10(6), 78 (1974).
Bahadori, M.N. and Edlin, F.E., Improvement of solar stills by the surface treatment of the glass, *Solar Energy*, 14(3), 339 (1973).
Bahari, E., Desalination processes and costs, *Chemical and Process Engineering*, 50, 71-75 (March, 1969).
Baibutaev, K.B. and Yakubov, U.N., Receiving of sweet water from the soil by condensation on transparent film under solar radiation action, *Applied Solar Energy (Geliotekhnika)*, 4(3), 40 (1968).
Baibutaev, K.B. and Achilov, B.J., Comparison tests on a solar distiller, *Applied Solar Energy (Geliotekhnika)*, 4(5), 50 (1968).
Baibutaev, K.B., Achilov, K.B. and Kamaeva, B.M., Experimental studies of the output of a solar still as a function of the ion sum of mineral matter in water, *Proc. All-Union Conference on Solar Energy Utilization*, Section 3-4, Erevan (1969). In Russian.
Baibutaev, K.B. and Astivov, B.M., On distilling water from the ground by helioinstallation, *Applied Solar Energy (Geliotekhnika)*, 5(4), 60 (1969).
Baibutaev, K.B., Achilov, B.M. and Kamaeva, G., Effect of salt concentration on the evaporation process in solar stills, *Applied Solar Energy (Geliotekhnika)*, 6(2), 83 (1970).

Baibutaev, K.B., Achilov, B.M. and Kamajeva, G., The influence of the salt content of water on evaporation in a solar still, *Applied Solar Energy (Geliotekhnika)*, 6(2), 110 (1970).

Baibutaev, K.B. and Achilov, B.M., Effect of the inclination of the transparent solar still on the condensation and collection processes, *Applied Solar Energy (Geliotekhnika)*, 6(3), 34 (1970).

Bairamov, R., Study of salt water desalination using very simple solar stills, Author's summary of Candidate Dissertation, Moscow (1963). In Russian.

Bairamov, R., Comparative tests of solar stills (In Russian), *Izvestia of Turkmenian Academy of Sciences* (1), 38-44 (1964).

Bairamov, R. and Atayev, Ya., *Economic Effect of Solar Stills* (In Russian) (5) (1966).

Bairamov, R.B., Rybakova, L.E. and Khamadov, A., Heat and mass transfer in models of solar stills of various configuration, *Applied Solar Energy (Geliotekhnika)*, 4(4), 77 (1968).

Bairamov, R.B. and Ataev, Ya., Technic and economic comparison of pastures watering methods in the Turkmen SSR (In Russian), *Geliotekhnika* 4 (5), 61-68 (1968).

Bairamov, R.B., Ataev, Ya. et al., Reports to All-Union Conference on the use of solar energy, *Erevan*, Vol. S-4, 45 (1969). In Russian.

Bairamov, R.B. and Annaev, M., Solar stills for industrial purposes in Turkmenia, Papers read to All-Union Conference on Solar Energy Utilization, Erevan (1969). In Russian.

Bairamov, R.B., A study of the distillation of water using solar energy in the Turkmenian SSR, Abstract of Doctoral Dissertation, Baku (1970). In Russian.

Bairamov, R. and Ataev, Ya., Results of a year-long solar still test, *Applied Solar Energy (Geliotekhnika)*, 7(1), 14 (1971).

Bairamov, R., Golubkov, B.N., Muradov, Dg. and Rachmanov, M., Experimental investigations of regenerative solar still installations under field conditions, *Applied Solar Energy (Geliotekhnika)*, 7(6), 94 (1971).

Bairamov, R.B. and Khamadov, A., Study of heat and mass exchange in plane-type solar distillers, *Applied Solar Energy (Geliotekhnika)*, 10(5), 24 (1974).

Bairamov, R., Seiitkurbanov, S., Agadzhanov, V. and Amanov, Ch., Procedure for determining load on watering sites and basic parameters of a distilling installation, *Applied Solar Energy*, 11(1), 15 (1975).

Barasoain, J.A. and Fontan, L., First experiment in the solar distillation of water in Spain (In Spanish), *Revista de Ciencia Aplicada, Madrid*, 72, 7-17 (Jan.-Feb. 1960).

Barduhn, O., Morse, R.N. et al., Review of the Second European Symposium on Fresh Water from the Sea, May 9-12, 1967. Athens, Greece, European Federation of Chemical Engineering, *Desalination*, 2(1), 5-12 (1967).

Bartali et al., Chimney and heated head solar still, *Heliotechnique and Development*, II, 431 (1976).

Battelle Memorial Institute (Landry, B.A., Eibling, J.A., and Thomas, R.E.), *Research Investigations of Multiple-effect Evaporation of Saline Waters by Steam from Solar Radiation*, OSW Report No. 2, PB 161377, 58 pages (December, 1953).

Battelle Memorial Institute (Bloemer, J.W., Collins, R.A., and Eibling, J.A.), *Study and Field Evaluation of Solar Sea-water Stills* (June 1960), OSW Report No. 50, PB 171934, 123 pages (September, 1961).

Battelle Memorial Institute (Bloemer, J.W., Irwin, J.R. and Eibling, J.A.), *Design of a Basin-type Solar Still* (May 1964), OSW Report No. 112, PB 181697, 26 pages (June 1964).

Battelle Memorial Institute (Bloemer, J.W., Irwin, J.R. and Eibling, J.A.), *Second Two Years' Progress on Study and Field Evaluation of Solar Sea-water Stills* (Sept. 1964), OSW Report No. 147, 83 pages (July 1965).

Selected Bibliography

Battelle Memorial Institute (Bloemer, J.W., Irwin, J.R. and Eibling, J.A.), *Final Three Years' Progress on Study and Field Evaluation of Solar Seawater Stills* (June, 1965), OSW Report No. 190, 87 pages (May 1966).

Baum, V.A., Prospects for the applications of solar energy, and some research results in the U.S.S.R., *Proc. World Symp. on Applied Solar Energy,* Phoenix, Ariz., 289-298 (Nov. 1955).

Baum, V.A., Solar radiation: Experience, possibilities and prospects of its exploitation, *Proc. New Delhi Symposium on Wind and Solar Energy,* UNESCO/NS/AZ/191/ANNEXE 23, Paris, 14 pages (1956).

Baum, V.A. (Editor), *Utilization of Solar Energy* (In Russian), Akademiya Nauk S.S.S.R., Moscow $\underline{1}$ (1957), $\underline{2}$ (1960). (Research and calculations regarding solar water stills, by P.M. Brdik, 136-150).

Baum, V.A., Technical characteristics of solar stills of the greenhouse type (In Russian), *Thermal Power Engineering, Utilization of Solar Energy,* $\underline{2}$, Academy of Sciences, USSR, Moscow, 122-132 (1960).

Baum, V.A., Solar distillers, United Nations Conference on New Sources of Energy, Paper 35/S/119, Rome, 43 pages (August 1961).

Baum, V.A. and Bairamov, R., Heat transfer in greenhouse type solar stills (In Russian), *Izv. AN Turkm. SSR, ser. fiz - tekh., khim. i geol. nauk,* (3) (1963).

Baum, V.A. and Bairamov, R., Heat transfer in hothouse type solar stills, *Ser. FTKh i GN,* No. 3 (1965).

Baum, V.A. and Bairamov, R., Heat and mass transfer processes in solar stills of hotbox type, *Solar Energy,* $\underline{8}$(3), 78-82 (1964).

Baum, V.A. and Bairamov, R., Prospects of solar stills in Turkmenia, *Solar Energy,* 10(1), 38-40 (1966).

Baum, V.A., Bairamov, R. and Khambadov, A., Solar hotbed-type distiller yield, *Geliotekhnika, Acad. of Sciences (USSR),* $\underline{3}$(3), 47-50 (1967).

Baum, V.A., Bairamov, R. and Toiliyev, K., Calculation method of solar distiller technical characteristics with consideration of nonstationary work (Nonsteady state), *Applied Solar Energy (Geliotekhnika),* $\underline{3}$(5), 52-58 (1967).

Baum, V., Bairamov, R. and Malevski, Yu., Possibilities of using solar energy in reclaiming deserts, *Problems of Desert Reclamation,* Ashkhabad (5) (1967).

Baum, V.A., Possibilities of the use of solar stills in arid zones of the world, Paper prepared for U.N. Solar Distillation Panel Meeting, 35 pages (Oct. 14-18, 1968).

Baum, V.A., Solar energy developments in USSR, *The Nucleus (Journal of the Pakistan AEC),* $\underline{5}$ (1 and 2), 11-13 (Jan.-June 1968).

Baum, V.A., Bayaramov, R.B. and Malevsky, V.M., The solar still in the desert, *Proc. International Solar Energy Congress,* Melbourne, p.426 (1970).

Baum, V.A., Bairamov, R.B., Annaev, M. and Rybakova, L.E., Experiments with a solar still in Karakum Desert, *Proc. First All-Union Scientific-Technical Conference on Replenishable Sources of Energy,* No. 1, Euergiya (1972). In Russian.

Baum, V.A., Technological characteristics of solar greenhouse stills, *Solar Energy Utilization,* $\underline{2}$, AN SSSR, Moscow (1980). In Russian.

Bayramov, R., A study of water distillation by means of the simplest solar stills (Dissertation in Russian), Power Engr. Inst. of the Acad. of Sciences, USSR, Moscow (1963).

Belloni, A., New system of solar distillation (In Italian), *Rev. Marittima, Supp. Tech.* 223-36 (July 1948).

Belmonte, E., Mejoramiento del destilador solar USM-5 y la comparacion con el modelo USM-4 para una planta de 500 m^2 en Quillaqua, Professional Thesis for the Degree of Mechanical Engineer, Federico Santa Maria Technical University, Valparaiso, Chile (1970).

Bennett, Iven, Monthly maps of mean daily insolation for the United States, *Solar Energy*, 9(3), 145-158 (1965).

Bettaque, Rolf, The Saltwater greenhouse. A technology for the use of saltwater for the growing of plants under controlled conditions, *Proc. ISES, New Delhi, India*, 3, Series 51, Abstract No. 1134, 1667 (1978).

Bhushan, Bharat, Assessment of the appropriateness of solar desalination technology for rural areas in arid regions of India, Mimeograph, 19th April (1978).

Bjorksten Research Laboratories, Inc., *Development of Plastic Solar Stills*, Office of Saline Water Report No. 24, PB 161398, 29 pages (Feb. 1959).

Bjorksten Research Laboratories, Inc., *Weathering Tests of Plastics and Design of Suspended Envelope Stills*, OSW Report No. 30, PB 161095, 40 pages (Sept. 1959).

Blanco, P., 'RM' multistage distiller (in French), *C.O.M.P.L.E.S. (Cooperation Mediterraneene pour l'Energie Solaire, Marseilles*, Bulletin No. 8, 66-69 (May 1965).

Blanco, P., Gomella, C. and Barasoain, J.A. 'Las Marinas' Solar conversion pilot plant (Spain), *Proc. First International Symposium on Water Desalination*, Washington, D.C. (October, 1965), 2, 817-824 (1967).

Blanco, P., Gomella, C. and Barasoain, J.A., Solar potabilization pilot plant at Las Marinas, Spain (In French), *C.O.M.P.L.E.S., Marseilles*, Bulletin No. 9, 29-41 (December 1965).

Blanco, P., Gomella, C., *Preliminary Report about the Pilot Solar Distillation Plant of Las Marinas* (In French), National Commission for Special Energies, Madrid, 4-5 (May 1966).

Blanco, P., Gomella, C. and Barasoain, J.A., Installation of sea water and brackish water potabilization plant at Las Marinas, Spain (In French), *C.O.M.P.L.E.S., Marseilles*, Bulletin No. 11, 32-38 (December 1966).

Blanco, P., *Final Report about the Transactions of the International Solar Distillation Research Center* In French), National Commission for Special Energies, Madrid, Spain, 25 pages (June 1967).

Blanco, P., Gomella, C. and Barasoain, J.A., Solar distiller project for the Nueva Tabarca Island, Alicante, Spain (In French), *C.O.M.P.L.E.S. Marseilles*, Bulletin No. 12, 47-55 (July 1967).

Blanco, P., Personal communication (Feb. 27, July 10 and Sept. 3, 1969).

Bloemer, J.W., Collins, R.A. and Eibling, J.A., Field evaluation of solar sea water stills, *Saline Water Conversion, Advances in Chemistry*, Series No. 28, American Chemical Society, Washington, D.C., 166-177 (1960).

Bloemer, J.W. and Eibling, J.A., *A Progress Report on Evaluation of Solar Sea-water Stills*, ASME paper No. 61-WA-296 (Nov. 1961).

Bloemer, J.W., Collins, R.A. and Eibling, J.A., *Study and Evaluation of Solar Sea-water Stills*, Saline Water Research and Development Progress Report No. 50, Office of Saline Water, Washington (1961).

Bloemer, J.W., Experimental investigation of the effect of several factors on solar still performance, M.Sc. Thesis, The Ohio State University, 40 pages (1963).

Bloemer, J.W., Eibling, J.A., Irwin, J.R. and Lof, G.O.G., *Analog Computer Simulation of Solar Still Operation*, ASME paper 63-WA-313, 8 pages (Nov. 1963).

Bloemer, J.W., Eibling, J.A., Irwin, J.R. and Lof, G.O.G., Solar distillation — A review of Battelle experience, *Proc. First International Symposium on Water Desalination*, Washington, D.C. (October, 1965), 2, 609-621 (1967).

Bloemer, J.W., Eibling, J.A. and Irwin, J.R., Solar distillation of sea water, *Batelle Technical Review*, 14(10), 14-18 (Oct. 1965).

Bloemer, J.W., Eibling, J.A., Irwin, J.R. and Lof, G.O.G., A practical basin-type solar still, *Solar Energy*, 9(4), 197-200 (Oct.-Dec. 1965).

Bloemer, J.W., *Factors affecting Solar-still Performance*, ASME Paper 65-WA/SOL-1, 8 pages (Nov. 1965).

Selected Bibliography

Bobrovnikov, G.N., Achilov, B.M., Kamilov, O.S., Kakharov, S., and Odinaev, A., Determining the thermal losses to the atmosphere from a solar still, *Applied Solar Energy*, 15(4), 66-68 (1979).

Bocic, Victor A. (Title unknown), Master's thesis, Technical University of Santa Maria, Valparaiso, Chile, 216 pages (1958).

Boelter, L.M.K., Gordon, H.S. and Griffin, J.R., Free evaporation into air of water from a free horizontal quiet surface, *Indus. Eng. Chem.*, 38(6), 596 (1946).

Boldrin, B. and Scalabrin, A continuous solar still with distinct heating and condensing partitions, *Heliotechnique and Development*, edited by M.A. Kettani and J.E. Sousson, Vol. II, 452, DAA (1976).

Boldrin, B., Scalabrin, G., Lazzarin, R. and Sovrano, M., A new continuous solar still, *Proc. ISES, New Delhi, India*, 3, Series 51, Abstract No. 161, 1502 (1978).

Boutaric, A. (Title unknown), *Recherches et Inventions*, 8, 205-215 (1927).

Boutaric, A., Solar heat and its use (In French), *Chaleur et Ind.*, 11, 59-66, 147-155 (1930).

Boutiere, H., Culture en zone aride et serres distillateurs solaires, *Cooperation Mediterraneene pour l'Energie Solaire*, Athens, Greece (Oct. 1971).

Boutiere, H., Culture en zone aride et serre-distillateurs solaires, *COMPLES Bulletin*, No. (1972).

Brace Research Institute, McGill University, How to make a solar still (plastic covered), *Do-It-Yourself Leaflet*, No. 1, 7 pages (January 1965).

Brancker, A.V., Demineralization of saline water, *Indus. Chemist*. 36(428), 489-496 (Oct. 1960).

Brancker, A.V., Solar stills in Iraq, *Nature*, 85, No. 4710 (1960).

Bratt, L.C., *Evaluation of the Mount Solar Still*, Stanford Research Institute, Project C-855, Technical Report No. 1 (Aug. 19, 1953).

Brice, D.B., Saline water conversion by flash evaporation utilizing solar energy, *Saline Water Conversion II, Advances in Chemistry* Series 38, American Chemical Society, Washington, D.C., 99-116 (1960).

Brice, D.B., Sunshine, fuel combine for desalination, *Chem. and Eng. News*, 40, 72 (April 2, 1962).

Bridlik, P.M., The condensation of water vapour from a water air mixture in solar desalinator. Candidate Dissertation (In Russian), Power Inst. of the USSR Academy of Sciences (1955).

Bridlik, P.M., Testing and designing solar desalination installations (In Russian) article in *The Utilization of Solar Energy* (1), edited by V.A. Baum, Academy of Sciences, USSR, Moscow, 136-150 (1957) (Translation by technical information center WPAFB, F-TS-9651/V (Sept. 1959), 179-200).

Bridlik, P.M., Testing and analysis of solar stills, *Use of Solar Energy*. Sb. 1 (1957).

Bridlik, P.M., Testing and design of solar stills, in Collection: *Utilization of Solar Energy*, Moscow, AS USSR, No. 1 (1957). In Russian.

Bridlik, P. Condensation of water vapor (In Russian), *Ingenerno-Fisitchesky Journal, Minsk*, 11(3) (1959).

Bromley, L.A., *Properties of Sea Water and its Concentrates up to 200 deg C*, Office of Saline Water Research and Development Progress Report No. 227 (December, 1966).

Cadwallader, E.A., Domestic desalination units, *Industrial and Engineering Chemistry*, 55 (March, 1962).

Cadwallader, Edgar A., *Family Size Solar Stills*, Office of Engineering, AID, 12 pages (January 1967).

Caufourier, P., Using solar heat through self-evaporation equipment (In French), *Le Genie Civil, Paris*, 82(14), 327-329 (April 1923).

Chhabra, A.K. and others, 5000 m^3 per day solar desalination plant — A case study, *Proc. National Solar Energy Convention*, Calcutta, India, pp.211 (1976).

Cillie, G.G., *Solar Distillation of Saline Water*, Water Research Div. Report, National Chemical Research Inst., C.S.I.R., Pretoria, South Africa, 26 pages (Oct. 1955).

Clark, Blake, Fresh water from the sea, *Pop.Sci.*, 172(3), 117-120, 266 (Mar. 1958).

Clarke, G.L. and James, H.R., Laboratory analysis of the selective absorption of light by sea water, *J. Opt. Soc. Am.* 29, 43 (1939).

Coanda, H., Changing salt water into soft water (In French), *Science et Vie*, D. Vincendon, Paris, 86(443), 146-151 (August 1954).

Coffey, James P., Vertical solar distillation, *Solar Energy*, 17, 373 (1975).

Coo, Raquel Wilson, Personal communication to Julio R. Hirschmann (June 22, 1961).

Cooper, P.I. and Appleyard, J.A., The construction and performance of a three effect, wick type, tilted solar still, *Sun at Work*, 12(1), 4-8 (First Quarter, 1967).

Cooper, P.I., Digital simulation of transient solar still processes, *Solar Energy*, 12(3), 313-331 (May 1969).

Cooper, P.I., The absorption of radiation in solar stills, *Solar Energy*, 12(3), 333-346 (May 1969).

Cooper, P.I., Solar distillation, *Solar Energy Progress in Australia and New Zealand*, Publication of the Australian and New Zealand Section of the Solar Energy Society (8), 45 (July 1969).

Cooper, P.I., Design and operational experience with CSIRO MK VI solar stills, Presented at AIRAH-SES Meeting, Perth (May 1971).

Cooper, P.I., Some factors affecting the absorption of solar radiation in solar still, *Solar Energy*, 13, 373 (1972).

Cooper, P.I., Digital simulation of experimental solar still data, *Solar Energy*, 14, 451 (1973).

Cooper, P.I., Maximum efficiency of a single effect solar stills, *Solar Energy*, 15, 205 (1973).

Cooper, P.I. and Read, W.R.W., Design philosophy and operating experience for Australian solar stills, *Solar Energy*, 16, 1 (1974).

C.S.I.R.O., *How to Make a Solar Still*, Melbourne, Australia, Diagrams, Steps 1-8, 1 page (1954).

C.S.I.R.O. (Australia), *Solar Energy Progress in Australia and New Zealand* (5), 9-10 (July 1966).

C.S.I.R.O. (Australia), 16-mm color film, Water from the Sun - the Coober Pedy solar still, showing construction details, 15 minutes (1967).

C.S.I.R.O. (Australia), *Desalting Water in Rural Areas*, Rural Research in C.S.I.R.O. (58), 8 pages (March 1967).

Daniel, F. and Duffie, J.A. (Editors), *Solar Energy Research*, University of Wisconsin Press, Madison, Wisconsin, 290 pages (1955). (Solar Distillation (Everett D. Howe), pp.107-109, Distillation with Solar Energy (Maria Telkes), pp.111-118; Solar Distillation of Sea Water in Plastic Tubes (Jeremiah T. Herlihy and Farrington Daniels), pp.125-126; A Survey of U.S. Patents Pertaining to Utilization of Solar Energy (John A Duffie), pp.255-265).

Daniels, Farrington, *Possibilities for Solar Distillation on South Pacific Islands*, Report to the South Pacific Science Board of U.S. Nat'l. Acad. of Sciences - Nat'l Research Council by University of Wisconsin, 32 pages (March 8, 1963).

Daniels, Farrington, *Direct Use of the Sun's energy*, Yale University Press, New Haven, 374 pp (1964). (Chap. 10 Distillation of Water, pp.167-195).

Daniels, Farrington, Construction and tests of small solar stills, *Proc. Solar Energy Society Annual Meeting*, Phoenix, Arizona, 18 pages (March, 1965).

Daniels, Farrington, Direct use of the Sun's energy, *American Scientist*, 55(1), 15-47 (March 1967).

Selected Bibliography

Daniels, Farrington, Solar energy use in family-sized units, *The Nucleus (Journal of the Pakistan AEC)* 5 (1 and 2), 4-10 (Jan-June 1968).

Daniels, F., Tests of small plastic water stills, Presented at the 1970 International Solar Energy Society Conference, Paper No. 5/115, Australia (Mar. 1970).

Datta, R.L., Gomkale, S.D., Ahmed, S.Y. and Datar, D.S. (India), Evaporation of Sea water in solar stills and its development for desalination, *Proc. First International Symposium on Water Desalination*, Washington, D.C., October, 1965, 2, 193-199 (1967).

Datta, R.L., Personal communication (July 31, 1969).

Davis, J.E., Lof, G.O.G. and Duffie, J.A., *Energy and Mass Transfer in Partial-pressure Distillation*, OSW Report 131, PB 166408 (1964).

Davison, R.R., Harris, W.B. and Moor, D.H., Coupled solar still, solar heater, *Proc. Int. Symp. on Fresh Water from the Sea*, 2, 437 (1976).

DeJong, T., An introduction to solar distillation, *Proc. American Society of Civil Engineers, Journal of the Sanitary Engineering Division*, SA4, Paper 1704, 84, 40 pp. (July 1958).

Della Porta, G.B., *Magiae naturalis libri XX*, Napoli (1589).

Della Porta, G.B., *De distillatione libri IX*, Roma (1608).

Delyannis, A.A., Solar stills provide an island's inhabitants with water, *Sun at Work*, 10(1), 6-8 (First Quarter, 1965).

Delyannis, A. and Piperoglou, E., Solar distillation in Greece, *Proc. First International Symposium on Water Desalination*, Washington, D.C., October 1965, 2, 627-633 (1967).

Delyannis, A., Sea water conversion developments in Greece, Paper at the International Symp. on Desalination, Milano (Apr. 18-20, 1966).

Delyannis, A. and Piperoglou, E., Solar distillation developments in Greece, *Sun at Work*, 12(1), 14-18 (1967).

Delyannis, A., Water from the Sun, *New Scientist*, 34, 388-389 (May 18, 1967).

Delyannis, A. and Piperoglou, E., Actual and potential desalination projects in Greece, *Acqua Dolce dal Mare, Milano*, 107-116 (1967).

Delyannis, A. and Piperoglou, E., The Patmos solar distillation plant, Techn. note, *Solar Energy*, 12, 113-115 (1968), (Also *Scientia*, No. 133, 39-44 (1967)).

Delyannis, A. and Delyannis, E., Operation of solar stills, Chapter prepared for U.N. Solar Distillation Panel Meeting, 5 pages (Oct. 11-18, 1968).

Delyannis, A. and Delyannis, E., The Gwadar, Pakistan, Solar Distillation Plant, Solar Energy Soc. Meeting, Palo Alto, California, 8 pages (Oct. 1968).

Delyannis, A. and Delyannis, E., Sea water distillation through solar radiation (In German), *Chemie-Ingenieur-Technik*, 41(3), 90-96 (Feb. 1969).

Delyannis, A. and Delyannis, E., *Handbook of Saline Water Conversion Bibliography*, Technical University, Athens, Greece; Antiquity - 1940, 1 (1967); 1941-1950, 2 (1967); 1951-1954, 3 (1968); 1955-1956, 4 (1968); 1957-1958, 5 (1968); 1959-1960, 6 (1968); 1961-1962, 7 (1969).

Delyannis, A. and Delyannis, E., Solar desalting, *Chemical Engineering*, 77(23), 136 (1970).

Delyannis, A. and Howe, E.D., *Report of Working Party on Recommended Procedure for Costing of Solar Stills*, C.S.I.R.O.; Division of Mechanical Engineering, Melbourne, Internal Report No. 77, 17 (1977).

Delyannis, A. and Delyannis, E., Solar distillation plant of high capacity, *Proc. 4th Int. Symp. on Fresh Water from Sea*, 4, 487 (1973).

Delyannis, E. and Delyannis, A., Solar applications in desalting, *Desalination*, 23, 541 (1977).

Delyannis, E. and Delyannis, A., Water desalination, *Naturwissenschaften*, 65, 462 (1978).

Dixit, D.K. and Deshmukh, S.T., Solar energy utilization for desalination, *Proc. National Solar Energy Convention*, Calcutta, India, pp.207 (1976).

Duffie, John A., A survey of U.S. patents pertaining to utilization of solar energy, in *Solar Energy Research*, edited by F. Daniels and J.A. Duffie, University of Wisconsin Press, Madison, Wisc., pp.255-265 (1955).

Duffie, John A., New materials in solar energy utilization, General Report at U.N. Conference on New Sources of Energy, Paper 35/GR/12(S), Rome, 18 pages (Aug. 1961).

Dunkle, R.V., Solar water distillation: The roof type still and a multiple effect diffusion still, *International Developments in Heat Transfer*, A.S.M.E., 895-902, Part 5 (August 28-Sept. 1, 1961). (Also CSIRO, Victoria, Australia, publ. 108 (1961)).

Dunkle, R.V., A simple solar water heater and still, Presented at joint AIRAH-SES Meeting, Perth (May 1971).

Du Pont de Nemours & Co., *Tedlar, Polyvinyl Fluoride Film*, Technical Information Bulletins, Nos. TD, Wilmington, Delaware.

Du Pont de Nemours & Co., *Design, Construction, and Operation of an Inflatable Plastic Solar Still*, Report to OSW, 30 pages (March, 1958).

Dzhubalieva, P.A., Determination of the aerodynamic coefficients of solar stills in relation to the leakage of steam-air mixture, *Applied Solar Energy* (trans. of *Geliotekhnika*), $\underline{1}$(4), 31-36 (July-August, 1965).

Dzhubalieva, P.A., Effect of leakage from a solar still on its performance under thermal head, *Applied Solar Energy*, (trans. of *Geliotekhnika*), $\underline{1}$(4), 37-42 (July-Aug., 1965).

Eckstrom, R.M., Design and construction of the Symi still, *Sun at Work*, $\underline{10}$(1), 7 (First Quarter, 1965).

Eckstrom, R.M., Field development and construction of plastic solar stills, Paper at Solar Energy Society meeting, Boston (Mar. 21-23, 1966).

Eckstrom, R.M., New solar stills in Greek Islands, *Sun at Work*, $\underline{11}$(2), 13 (Second Quarter, 1966).

Eckstrom, R.M., Personal communication (May 26, 1969).

Edlin, F.E., New data for the use of solar energy (In French), *Colloque International sur les Applications Thermiques de L'energie Solaire dans le Domaine de la Recherche et de L'industrie*, Montlouis, France (June 1958).

Edlin, F. (Title unknown), *Du Pont Magazine*, $\underline{53}$(2) (April-May, 1959).

Edlin, F.E. and Willauer, D.E., Plastic films for solar energy applications, United Nations Conference on New Sources of Energy, Paper 35/S/33, Rome 28 pages (Aug. 1961).

Edlin, Frank, A report on world solar activities, *Sun at Work*, 12-13 (Second Quarter, 1967).

Edlin, F.E., Water, food and solar energy, *Proc. Energy and Power Conference*, Rockefeller Foundation, New York, 19 pages (July 1967).

Edlin, Frank E., Worldwide progress in solar energy, Inter-society Energy Conversion Engineering Conference, 1968 Record, Boulder, Colorado, 92-97 (Aug. 13-17, 1968).

Edlin, Frank E., Personal communication (Mar. 1, 1969).

Edson, L., McCracken, H. and Weldy, J., *How to Build a Solar Still*, Sea Water Conversion Laboratory, University of California, Series 75, Issue 13 (April, 1959). (Revised in 1962 and again in June, 1966 by B.W. Tleimat, 13 pages).

Eibling, J.A., Thomas, R.E. and Landry, B.A., An investigation of multiple-effect evaporation of saline waters by steam from solar radiation, Saline Water Conversion Program, Department of the Interior, 1953, Washington, D.C. (Research and Development Progress Report, 2).

Eibling, James A., Realities of solar energy, *Battelle Technical Review* (March 1957).

Eibling, J.A., Talbert, S.G. and Lof, G.O.G., Solar stills for community use - Digest of technology, *Solar Energy*, $\underline{13}$, 263 (1971).

Ellis, Cecil B., *Fresh Water from the Ocean*, Ronald Press, New York (1954), Chap. 8, Distillation by new heat sources, 161-168.

El-Nashar, Ali, M., Design aspects of a solar assisted reverse Osmosis desalting unit for urban communities, *Proc. Int. Cong. on Desalination and Water Reuse*, 1979, Published by Elsevier Scientific Publications, Amsterdam, Netherlands and NY, 32, 239 (1980).

Erb, R.A., *Producing Permanently Hydrophylic Surfaces on Plastic Films for Solar Stills*, OSW Report No. 29, PB 161064 27 pages (Sept. 1959).

Erb, R.A., Thelen, E. and Jackson, F.L., *Producing Selectively Infrared-reflecting Surfaces on Plastic Films for Solar Stills*, OSW Report No. 53, PB 181035, 33 pages (October 1961).

Fabuss, B.M. and Korosi, A., *Thermophysical Properties of Saline Water Systems*, Office of Saline Water Research and Developments Progress Report No. 189 (May 1966).

Falvey, Henry, T. and Todd, Clement J., Cencentric tube solar still, *Proc. Joint Conference - Sharing the Sun - of the ISES*, American Section and Solar Energy Society of Canada, Winnipeg, Manit, 5, 210 (1976).

Fitzmaurice, R. and Seligman, A.C., Some experiments on solar distillation of sea water in Cyprus during the summers of 1954 and 1955, *Trans. Conference on the Use of Solar Energy*, Tucson, Arizona, 1955, University of Arizona Press, 3, 109-118 (1958).

Fontan, L. and Barasoaín, J.A., Some experiments concerning the solar distillation of water on a small scale (In French), U.N. Conference on New Sources of Energy, Paper 35/S/73, Rome, 20 pages (August 1961).

Fraser, J.H. and Davis, H.F., Laboratory investigation of concentrating industrial wastes by freeze crystallization, AICHE, 79th National Meeting, Houston, Texas (March 16-20, 1975).

Freilander, R., Large-area solar distillation plant and sprinkler installation, Fed. Rep. of Germany patent 2, 626, 902/A (22 Dec., 1977). In German.

Frick, G., *Destiladores solares desarrollados en el IDIEM, Santiago (1956-1957)*. Centro Universitario Zona Norte, Universidad de Chile, 1, (1958).

Frick, G., *Destiladores solares*, Departmento de Investigaciones Cientificas, Antofagasta (DICA) Centro Universitario Zona Norte, Universidad de Chile, 3 (1960-1961).

Frick, G.P., Some new considerations about solar stills, International SES Conference, Melbourne paper 5/78 (1970).

Frick, G. and Hirschmann, J., Theory and experience with solar stills in Chile, *Solar Energy*, 14(4), 405-413 (1973).

Frick, G. and Sommerfeld, J.V., Solar stills of inclined evaporating cloth, *Solar Energy*, 14(4), 427-431 (1973).

Frick, German, *Studies on the Utilization of Solar Energy in Stills and Cookers* (In Spanish), Centro Universitario Zona Norte, Universidad de Chile, 94-102 (1958).

Furon, R. (translated by P. Barnes), *The Problem of Water: A World Study*, American Elsevier Publishing Co., New York, 90 pages (1967).

Garg, H.P. and Krishnan, A., Determination of optimum orientation of solar stills and results of an experimental study at Jodhpur, *Proc. 5th Meeting All India Solar Energy Working Group and Conference*, Madras, 85 (Nov. 1973).

Garg, H.P., Design studies on conventional double sloped solar stills, *Indian and Eastern Engineers*, 117(8), 357 (1975).

Garg, H.P. and Thanvi, K.P., Development of a new floating tray solar still, *Proc. National Solar Energy Convention*, Calcutta, India, pp.209 (1976).

Garg, H.P. and Mann, H.S., Effect of climatic, operational and design parameters on the year round performance of single sloped and double-sloped solar still under Indian and arid zone conditions, *Solar Energy*, 18, 159 (1976).

Garg, S.K. and Datta, R.L., Humidification-dehumidification technique for sea water desalination, presented at Conference on Research and Industry, New Delhi (Dec. 20-21, 1965).

Garg, S.K., Gomkale, S.D. and Datta, R.L., Use of solar energy for production and supply of water from salt water, presented at Symp. on Community Supply and Waste Disposal, at CPHERI, Nagpur, India (Dec. 19-20, 1966).

Garg, S.K., Mehta, M.H. and Datta, R.L., Conversion of sea water into fresh water for boilers and other industries, Seminar on Water Treatment, Bombay, July (1967).

Garg, S.K., Gomkale, S.D., Datta, R.L. and Datar, D.S., Development of humidification-dehumidification technique for water desalination in arid zone of India, Second European Symp. on Fresh Water From the Sea, Athens (May 9-12, 1967). (Also in *Desalination*, $\underline{5}$(1), 55-63 (1968)).

Garg, S.K. and Datta, R.L., Humidification-dehumidification technique of water desalination for various purposes using different sources of energy, paper at Solar Energy Society meeting, Palo Alto, California (Oct. 1968).

Garg, S.K., Mehta, M.H., Gomkale, S.D. and Datta, R.L., Experience in the operation of humidification-dehumidification pilot plant of sea water, *Desalination*, $\underline{6}$(5-6), 61 (1968).

Garg, S.K. and Datta, R.L., H-D technique of desalination for water supply in light houses, Annual General Meeting of Indian Institute of Chemical Engineers, Kanpur, Dec. (1971).

Garret, C.R. and Farber, E., *Performance of a Solar Still*, ASME Paper No. 61-SA-38, 8 pages (June 1961).

Gilliland, E.R., Fresh water for the future, *Ind. Eng. Chem.*, $\underline{47}$(12), 2410-2422 (Dec. 1955).

Ginestous, M. (Title unknown), *Bull. Doc. Gener. l'Agriculture* (1928).

Ginestous, M. (Title unknown), *Revue Agricole de l'Afrique du Nord* (1928).

Goghari, H.D. and Gomkale, S.D., Some aspects of civil engineering in water desalination, *Indian Chem. Manufr.*, $\underline{6}$(5-6), 58 (1968).

Goghari, H.D., Natu, G.L. and Gomkale, S.D., Problems in construction and maintenance of large solar stills plants, *Proc. National ISES Solar Energy Convention*, I.I.T. Bombay, Paper/44, 231 (1979).

Gomella, C. (Title unknown - summary of work in Algeria and Morocco), paper presented at meeting, Organisation Europeene de Cooperation Economique, Alger (May 3-5, 1955).

Gomella, C., The problem of the demineralization of brackish waters by solar distillation (In French), *Terres et Eaux*, $\underline{6}$(25), 2e Trimestre, 4-31 (1955).

Gomella, Cyril, Demineralization of saline waters by solar distillation (In French), *Terres et Eaux*, $\underline{6}$(26), 3e Trimestre, 10-16 (1955).

Gomella, C., Practical possibilities for the use of solar distillation in under-developed arid countries, *Trans. Conference on the Use of Solar Energy*, Tucson, Arizona, 1955, University of Arizona Press, $\underline{3}$, 119-133 (1958).

Gomella, C., Solar distillation developments in the Eastern Mediterranean, *Proc. Symposium on Saline Water Conversion*, National Academy of Sciences, National Research Council, Washington, D.C., Publication No. 568, 131-136 (Nov. 4-6, 1957).

Gomella, C., In *Saline Water Conversion: Proc. Symposium on Saline Water Conversion 1957*, p.131, National Academy of Sciences - National Research Council, Washington.

Gomella, C., Contribution to the study of solar distillation, (In French), *Colloque International sur les Applications Thermiques de l'energie Solaire dans le Domaine de la Rechercher et de L'industrie*, Mont. Louis (1958).

Gomella, C., Use of solar energy for the production of fresh water, U.N. Conference on New Sources of Energy, General Report Paper 35/GR/19(S), Rome, 42 pages (August 1961).

Gomella, C., Possibilities of extending the dimensions of solar distillers (In French), U.N. Conference on New Sources of Energy, Paper 35/S/107, Rome, 16 pages (August 1961).

Gomella, C., Principles governing the construction of large dimension distillers (In French), Second European Symposium on Fresh Water From the Sea, Athens (May 9-12, 1967), Published in *Desalination*, 4(1), 61-65 (1968).

Gomkale, S.D., Ahmed, S.Y., Datta, R.L. and Datar, D.S., Conservation of sea water by solar still, Presented at the Symposium on Salt and By-products, Bhavnagar (1964).

Gomkale, S.D., Ahmed, S.Y., Datta, R.L., Datar, D.S., Fresh water from sea by solar still, Paper presented at the Annual meeting of the Indian Institute of Chemical Engineers, Bangalore, India (Dec. 1964).

Gomkale, S.D., Ahmed, S.Y., Datta, R.L. and Datar, D.S., Fresh water from sea by solar stills, *Indian Chemical Engineering*, 7(4), 97 (1965).

Gomkale, S.D., Solar stills as source of drinking water for isolated communities and for distilled water production, *Indian Chem. Manufacturers*, 6(5-6), 53 (1968).

Gomkale, S.D., Review of the First Conference on Water Desalination in India, *Desalination*, 4(1), 131-134 (1968).

Gomkale, S.D. and Datta, R.L., Solar stills for water supply for lighthouse with a case study for Vengurla Rocks, *Chemical Age of India*, 19(6), 435-438 (June, 1968).

Gomkale, S.D., Garg, S.K. and Datta, R.L., Solar distillation techniques - their development and application in India, *Proc. Int. Solar Energy Society Conference*, Melbourne (1970).

Gomkale, S.D. and Datta, R.L., Some aspects of solar distillation for water purification, ISES Conference, Greenbelt, Maryland, May (1971).

Gomkale, S.D. and Datta, R.L., Some aspects of solar distillation for water purification, *Solar Energy*, 14(4), 387 (1973).

Gomkale, S.D., Datta, R.L., Solar distillation in India, *Annals of Arid Zone, Solar Radiation, India*, 15(3), 206 (1976).

Grange, B.W., The transmissivity of selected glasses and plastics to solar radiation, M.S. Thesis, University of California (May 1966).

Grune, Werner N. and Zandi, Iraj, Improved solar still process for desalting sea and Brackish waters, *Journal of American Water Works Association*, 52(8), 993-1005 (August 1960).

Grune, W.N., Thompson, T.L. and Collins, R.A., *New Applications of Thermodynamic Principles to Solar Distillation*, ASME Paper 61/SA/45, Los Angeles, California (June 15, 1961).

Grune, W.N., Thompson, T.L. and Collins, R.A., Forced convection multiple effect solar still for desalting sea and brackish waters, U.N. Conference on New Sources of Energy, Paper 35/S/14, Rome, 26 pages (August 1961).

Grune, W.N., Hughes, R.B. and Thompson, T.L., Operating experiences with natural and forced convection solar stills, *Water and Sewage Works*, 108, 378-383 (October 1961).

Grune, W.N., Hughes, R.B. and Thompson, T.L., Natural and forced convection solar stills, *Journal of the Sanitary Engineering Division, Proc. A.S.C.E.*, 88, SA 1, Paper No. 3036, 24 pages (January 1962). (Disc. by T. Dejong, 88, SA 5, Paper No. 3288, 47, 48 (Sept. 1962).

Grune, Warner N., Collins, Richard A., Hughes, Ross B. and Thompson, T. Lewis, *Development of an Improved Solar Still*, OSW Report No. 60, PB 181144, 115 pages (March 1962).

Gunaji, Narendra, N., Lunsford, Jesse V., Keyes, Conrad, G., Jr., Loth. William D. and Gregory, William S., *Disposal of Brine by Solar Evaporation*, Final Report - Phase 1 to OSW, Engineering Experiment Station, New Mexico State University, University Park, New Mexico, Tech. Report No. 34, 220 pages (Sept. 1966).

Gunaji, N.N. and Keys, C.G., Jr., *Disposal of Brine by Solar Evaporation*, Office of Saline Water Research and Development Progress Report No. 351, May 1968.

Hafez, M.M. and Elnest, M.K., Demineralization of saline water by solar radiation in the United Arab Republic, U.N. Conference on New Sources of Energy, Paper 35/S/63, Rome, 10 pages (August 1961).

Halacy, D.S., Jr., Solar Science Projects, Scholastic Book Service, New York (1963). (Solar still drawing shown in *Sun at Work* (Second Quarter, 1963)).

Hall, Raymond C., Theoretical calculations on the production of water from the atmosphere by absorption with subsequent recovery in a solar still, *Solar Energy*, 10(1), 41-45 (1966).

Hall, R.C., Rao, G.V. and Livington, R.J., Prediction of liquid hold up on an absorption tower for the production of water from the atmosphere with subsequent recovery in a solar still, *Solar Energy*, 12(2), 147-161 (1968).

Hamid, Y.H., An experimental solar still design for the Sudan, *Appropriate Technology*, 3(3), 12 (1976).

Harding, Josiah, Apparatus for solar distillation, *Proc. Institution of Civil Engineers*, 73, 284-288 (1883).

Hatzikakidis, Athan D., Evaporation by solar energy, *Sun at Work*, 12-14 (Fourth Quarter, 1964).

Hay, Harold R., New concepts in solar still design, *Proc. First International Symp. on Water Desalination*, Washington, D.C., Oct. 1965, 1, 511-527 (1967).

Hay, H.R., V-Cover solar stills, *Sun at Work*, 2(2), 6-9 (Second Quarter, 1966).

Hay, Harold R., Letter to Editor, *Sun at Work*, 19 (Fourth Quarter, 1966).

Hay, H.R., Plastic solar stills: Past, present, and future, *Solar Energy*, 14(4), 393-404 (1973).

Headley, O.St.C., Cascade solar still for distilled water production, *Solar Energy*, 15, 245 (1973).

Headley, O.St.C., Remarks on 'Cascade solar still for distilled water production', *Solar Energy*, 18, 75 (1976).

Headley, O.St.C., Augustine, St. and Morris, J.B., Design, construction, and operation of solar stills for producing distilled water, *Proc. Int. Conference on Solar Building Technology*, 2, 612 (1977).

Henneberry, G.O., Analysis of encrustation on orlon felt, Inter-departmental Memo, File 60/08, Brace Research Institute, McGill University (April 1, 1968).

Henrik, Weihe, Fresh water from sea water: Distilling by solar energy, *Solar Energy*, 13(4), 439 (1972).

Herlihy, J.T. (Title unknown), M.S. Thesis, University of Wisconsin (1953).

Herlihy, J.T. and Daniels, F., Solar distillation of sea water in plastic tubes, in *Solar Energy Research*, ed. by F. Daniels and J.A. Duffie, University of Wisconsin Press, Madison, Wisconsin, 125-126 (1955).

Heywood, J.B., *The Design of a Solar Water Still*, Cambridge University Engineering Society, 10 pages.

Hirschmann, Julio, Pilot plant to use solar energy in producing potable water from salt water with simultaneous generation of electric power (in Spanish), *Scientia, Valparaiso, Chile*, 25(4), 195-200 (December 1958).

Hirschmann, Julio R., Project of a solar energy pilot plant for the North of Chile, *Solar Energy*, 5(2), 37-43 (1961).

Hirschmann, J.R., Solar evaporation and distilling plant in Chile (in French), U.N. Conference on New Sources of Energy, Paper 35/S/23, Rome, 30 pages (August 1961).

Hirschmann, Julio, Project of a solar desalination plant for the North of Chile, *Proc. First International Symp. on Water Desalination*, Washington, D.C., Oct. 3-9, 1965, 1, 499-510 (1967).

Hirschmann, J., Evaporation solar y su application practica en Chile, *Scientia*, 136 (1968).
Hirschmann, J.R. and Roefler, S.K., Thermal inertia of solar stills and its influence on performance, *Proc. ISES*, Melbourne, 402 (1970).
Hirschmann, J.R., Solar distillation in Chile, *Desalination*, 17(1), 17-30 (1975).
Hirschmann, J., Solar distillation with evaporating wick, *Proc. Int. Symp. on Fresh Water from the Sea*, 2, 447 (1976).
Hirshleifer, J., De Haven, J.C. and Milliman, J.W., *Water Supply Economics, Technology and Policy*, University of Chicago Press, 378 pages (1960).
Hobbs, Wesley, *The Design and Construction of a Solar Heater for the Distillation of Water*, Haile Selassie 1 University, Dire Dawa, Ethiopia, Misc. Publication No. 7, 5 pages (June 1965).
Hodges, C.N. and Kassander, R., *Distillation of Saline Water utilizing Solar Energy in Multiple Effects System consisting of Separate Collector, Evaporator and Condenser*, Solar Energy Laboratory, University of Arizona, Introductory Report AD 275326, 20 pages (April 1, 1962).
Hodges, C.N., Thompson, T.L. and Groh, J.E., *Separate Component, Multiple Effect Solar Distillation*, Solar Energy Laboratory, University of Arizona, Interim Report No. 1, 48 pages (November 1962).
Hodges, C.N., *Separate Component, Multiple-effect Solar Distillation*, Solar Energy Laboratory, University of Arizona, Interim Report No. 2, 49 pages (March 1963).
Hodges, C.N., Thompson, T.L., Groh, J.E. and Sellers, W.W.D., The utilization of solar energy in a multiple-effect desalination system, *Journal of Applied Meteorology*, 3(5), 505-512 (October 1964).
Hodges, C.N., Groh, J.E. and Thompson, T.L., A report on the Puerto Penasco solar desalination plant, Paper presented at Solar Energy Society Annual Meeting (March 1965).
Hodges, Carl N., Groh, John E., Thompson, T. Lewis, Solar powered humidification cycle desalination (A report on the Puerto Penasco Pilot Desalting Plant), *Proc. First International Symp. on Water Desalination*, Washington, D.C., Oct. 1965, 2, 429-456 (1967).
Hodges, Carl N., Thompson, T. Lewis, Groh, John E. and Frieling, Donald H., *Final Report - Solar Distillation Utilizing Multiple-effect Humidification*, Solar Energy Research Laboratory, University of Arizona (January 31, 1966), OSW Report No. 194, 155 pages (May, 1966).
Hodges, Carl N., Future trends and unconventional applications, Chapter prepared for U.N. Solar Distillation panel meeting, 18 pages (Oct. 14-18, 1968).
Hollands, K.G.T., The regeneration of lithium chloride brine in a solar still, *Solar Energy*, 7(2), 39-43 (1963).
Hollingsworth, F.N., Solar energy to provide water supply for island, *Heat. and Vent.*, 45, 99 (Aug. 1948).
Hottel, H.C. and Woertz, B.B., The performance of flat-plate solar heat collectors, *Trans. ASME*, 91-104 (February 1942).
Howe, E.D., Fresh water from sea water, *Trans. Am. Geophy. Union*, 33(3), 417-422 (June 1952).
Howe, E.D., Sea water as a source of fresh water, *J. Amer. Water Works Assoc.*, 44(8), 690-700 (August 1952).
Howe, E.D., *Sea Water Research - a Technical and Economic Investigation of Certain Schemes for Producing Potable Fresh Water from Sea Water*, University of California, Dept. of Engr. Progress Reports (Sept. 1953 and Jan. 15, 1954).
Howe, E.D., Yuster, S.T., Hassler, G.L. and Tribus, Myron, *Summary Report for Conference of Saline Water Conversion Program*, University of California, Dept. of Engr. Publ. (April 20, 1954).
Howe, Everett D., Solar distillation, in *Solar Energy Research*, edit. by F. Daniels and J.A. Duffie, University of Wisc. Press, Madison, Wisconsin, 107-109 (1955).

Howe, E.D., Solar distillation, *Trans. Conference on the Use of Solar Energy - the Scientific Basis*, Tucson, 3, 159-169 (1955).

Howe, E.D., *Utilization of Sea Water*, University of California, Div. of Mech. Engr., Berkeley, California (1956).

Howe, E.D., Experiments under way to obtain demineralized water for cities, *Western City*, 32(10), 46-49 (Oct. 1956).

Howe, E.D., Progress in conversion of saline water, *J. American Water Works Assoc.*, 50, 319 (Mar. 1958).

Howe, E.D., Solar distillation experiments at the University of California, *Solar Energy*, 3(3), 31 (1959).

Howe, E.D. Use of solar energy for production of fresh water: small and large scale distillers, U.N. Conf. on New Sources of Energy, Rome (April 10, 1961).

Howe, Everett D., MacLeod, Lester H., Solar distillation research at the University of California, ASME paper 61/SA/37, 9 pages (June 1961).

Howe, E.D., Solar distillation research at the University of California, U.N. Conference on New Sources of Energy, Paper 35/S/29, Rome, 22 pages (August 1961).

Howe, Everett D., MacLeod, L.H. and Tleimat, B.W., Solar and atmospheric distillation, Saline Water Conversion Research, Berkeley Progress Report, University of California, Sea Water Conversion Lab., 11 pages (Dec. 31, 1962).

Howe, Everett D., The distillation of sea water and other low-temperature applications of solar energy, Chapter 12 in *Introduction to the Utilization of Solar Energy*, edited by Zarem and Duane D. Erway, McGraw-Hill Book Co., New York, 295-313 (1963).

Howe, Everett D., Solar still and inclined-tray solar still with copper-foil water basin, Descriptive pamphlet, Sea Water Conversion Lab., University of California, Richmond, California (Sept. 12, 1963).

Howe, E.D., Solar distillation on the pacific Atolls, *South Pacific Bulletin*, 14(2), 57-59 (April 1964).

Howe, E.D., Solar distillation problems in the developing countries, ASME Paper 64/WA/SOL-7, New York (Dec. 1964).

Howe, E.D., Pacific island water systems using combined solar still and rainfall collector, *Solar Energy*, 10(4), 175-181 (Oct.-Dec. 1966).

Howe, E.D. and Tleimat, B.W., Solar distillers for use on coral islands, Second European Symp. on Fresh Water from the Sea, Athens (May 9-12, 1967). Also in *Desalination*, 2(1), 109-115 (1967).

Howe, Everett D., Tleimat, Badawi W. and Laird, Alan D.K., *Solar Distillation*, University of California, Sea Water Conversion Laboratory, Report No. 67-2, Water Resources Center, Desalination Report No. 17, 52 pages (1967).

Howe, Everett D., Review of still types, Chapter prepared for U.N. Solar Distillation Panel Meeting, 34 pages (Oct. 14-18, 1968).

Howe, E.D. and Tleimat, B.W., Twenty years of work on solar distillation at the University of California, UNESCO Conference, The Sun in the Service of Mankind, Paris (July, 1973).

Howe, E.D., *Fundamental of Water Desalination*, Marcel Dekkar, Inc., NY (1974).

Howe, E.D. and Tleimat, B.W., Twenty years of work on solar distillation at the University of California, *Solar Energy*, 16, 97 (1974).

Howe, E.D., Some comments on solar distillation, *Proc. International Solar Energy Congress*, Melbourne, Australia (Extracts from the Conference bulletin No. 2) (1979).

Hummel, R.L. and Rudd, D.F., *A Solar Distillation Design for the Economic Production of Fresh Water from Solar Water (and Addenda)*, University of Michigan, 29 pages (1960), (Addenda, 13 pages).

Hummel, R.L., New concepts in solar stills leading to low cost fresh water from sea water, American Chemistry Society Meeting, St. Louis, Missouri, 23 pages (March 1961).

Selected Bibliography

Hummel, Richard L., Power as a by-product of competitive solar distillation, U.N. Conf. on New Sources of Energy, Paper 35/S/15, Rome, 22 pages (Aug. 1961).

Hummel, Richard L., A large scale, low cost, solar heat collector and its application to sea water conversion, U.N. Conference on New Sources of Energy, Paper 35/S/28, Rome, 19 pages (August 1961).

Hundemann, Audrey S., *Solar Distillation* (citation from the engineering index data base), NTI Search NTIS/PS-79/0033/SEES, Search period covered: 1970–Dec. 1978. Publ. by NTIS, Springfield (1979).

Irwin, J.R. and Fischer, R.D., *Phase I Summary Report on the Use of Low-cost Plastic Films in Solar Stills*, Battelle Memorial Inst., Report to OSW, 15 pages (Oct. 16, 1967).

Itty, P.I., Modes of energy transfer in a solar still, *Proc. National Solar Energy Convention*, I.I.T. Bombay, Paper/46, 242 (1979).

Jackson, R.D. and Van Bavel, C.H.M., Solar distillation of water from soil and plant material, *Science*, $\underline{149}$(3690), 1377 (Sept. 17, 1965).

Jackson, R.D. and Van Bavel, C.H.M., A simple desert survival still, *Sun at Work*, 4–6 (Fourth Quarter, 1966).

Jenkins, D.S., Sieveka, E.H., Saline water conversion, Paper presented at the Western Area Development Conference, Phoenix, Ariz., 28 pages (Nov. 1956).

Jenkins, D.S., Fresh water from salt, *Sci. Am.* $\underline{196}$(3), 37–45 (Mar. 1957).

Jenkins, D.S., Developments in saline water conversion, *J. Am. Water Works Assn.* $\underline{49}$(8), 1007–1019 (1957).

Jensen, Jean Smith (editor) *Applied Solar Energy Research – a Directory of World Activities and Bibliography of Significant Literature*, 2nd edition, Association of Applied Solar Energy (1959).

Jensen, Jean Smith, Harnessing the sun around the world, *The Sun at Work*, $\underline{4}$(1), 3–7 (March 1959).

Jensen, M.H., The use of waste heat in agriculture, *Proc. of the National Conference on Waste Heat Utilization*, Gattingurg, Tennessee (Oct. 1979).

J.R.M., Solar desalting gains acceptance in Greece, *Chemical Engineering*, $\underline{66}$, 42 (Dec. 20, 1965).

Kalloy, W., *Handbook of Saline Water Conversion Data*, U.S. Department of Interior Report, Washington, D.C.

Kamal, Ismat, *Prospectus of Desalination of Sea Water on the Makran Coast (West Pakistan)*, Pakistan Atomic Energy Commission (April 1967).

Karim, Munawar, Comments on the performance of solar stills, *Solar Energy*, $\underline{20}$, 361 (1978).

Kausch, O., *Direct Use of Solar Energy* (In German), Source unknown, 'Early Attempts at Solar Distillation Summarized' (Telkes).

Keller, E., Solar energy in distillation of sea water, *Ind. & Eng. Chem.*, $\underline{56}$, 10, 12 (Feb. 1964).

Kellong, M.W. Co. (New York), *Saline Water Conversion, Engineering Data Book*, U.S. Department of the Interior Charts 11:50 (July 1965).

Kendrew, W.G., *The Climates of the Continents*, 3rd edition, Oxford University Press, New York, 473 pages (1942).

Kettani, M.A. and Abdel-Aal, H.K., Production of magnesium chloride from the brines of desalination plants using solar energy, *Proc. 4th International Symposium on Fresh Water from the Sea*, Heidelbergh, Germany, $\underline{2}$, 509 (1973).

Kettani, M. Ali, Desalination by solar refrigeration, *Revue Internationale d'Heliotechnique*, p.39–44 (1976).

Kettani, M.A., Review of solar desalination, *Sun World*, $\underline{3}$(3), 76 (1979).

Khan, Ehsan Ullah, Practical devices for the utilization of solar energy, *Solar Energy*, $\underline{8}$(1), 17–22 (1964).

Khanna, M.L. and Mathur, K.N., Experiments on demineralization of water in North India, U.N. Conference on New Sources of Energy, Paper 35/S/115, Rome, 11 pages (August 1961).

Khanna, M.L., Solar water distillation in North India, *Proc. International Seminar on Solar and Aeolian Energy*, Sounion, Greece, September 1961, Plenum Press, New York, 59-71 (1964).
Khanna, Mohan Lal, Solar water distillation in North India, *J. Sci. and Indus Research*, 21A(9), 429-433 (Sept. 1962).
Khanna, M.L., The present status of demineralization of saline water, Symposium on Problems of Indian Arid Zone, Jodhpur, India (Nov. 23-Dec. 2, 1964).
Khanna, M.L. Problems of arid regions, *Sun at Work*, 10(2), 8-9 (1965).
Khanna, Mohan Lal, Symposium on arid zone problems - a review, *Sun at Work*, 11(1), 14 (1966).
Khanna, M.L., Solar still, *Encyclopaedic Dictionary of Physics*, Pergamon Press Ltd., London, 2, 332-336 (1967).
Khanna, M.L., How to use the present day technology of desalting of saline water in solving fresh-water shortages in India, *Chemical Age of India*, 19(6), 430-434 (1968).
Khatamov, S.O., Umarov, G.Ya, Tests of a plastic staircase-type solar still, *Applied Solar Energy (Geliotekhnika)*, 7(2), 68 (1971).
Kobayashi, M., A method of obtaining water in arid lands, *Solar Energy*, 7(3), 93-99 (1963).
Kokkaliaris, P.P., International research center for distillation, *Solar Energy*, 14(4), 423-425 (1973).
Kokkaliaris, P. (Greece), The solar distillation units in Greece, *Proc. International Solar Energy Congress*, Melbourne (Extracts from the Conference Bulletin No. 2) (1979).
Korpeev, N.R., Bairamov, R., Golubkov, B.N. and Rakhmanov, M., Corrosion prevention in solar stills with adiabatic evaporation, *Applied Solar Energy (Geliotekhnika)*, 8(5), 21 (1972).
Kumar, A., Physics of solar stills, Ph.D. thesis, Department of Physics and Centre of Energy Studies, I.I.T. Delhi, India (1981).
Kumar, U.N., Mittal, S.C. and Kakar, M.P., Development of solar still for the supply of laboratory distilled water and potable water for a typical village home, *Proc. ISES*, New Delhi, India, 3, Series 51, Abstract No. 138, 1492 (1978).
Kumar, U.N., Mittal, S.C. and Kakar, M.P., Design of a combined solar water heating-distillation system for rural application, *Proc. ISES*, New Delhi, India, 3, Series 51, Abstract No. 142, 1496 (1978).
Lanfagne, M., Private communication, Department of Agricultural Engineering, McGill University, Montreal, Canada (1971).
Lanfagne, M., Double stage solar stills, M.S. Thesis, Dept. Ag. Engg., McGill University, Montreal, Canada (1971).
Laooakam, R. and Bjorksten, J. Development of plastic solar stills for use in the large scale, low cost demineralization of saline water by solar evaporation, *Trans. Conf. on the Use of Solar Energy, The Scientific Basis*, Tucson, Arizona, 3, 99 (1955).
Lappala, Risto and Bjorksten, Johan, Development of plastic covered solar stills for use in the large-scale, low-cost demineralization of saline water by solar evaporation, *Trans. Conference on the Use of Solar Energy*, Tucson, Arizona, 3, 99-107 (1955).
La Parola, G., Apparatus for the distillation of brackish or sea water by means of solar energy (in Italian), *Notiziario Economico della Cirenaica*, Bengasi, Libya, 12 pages (Nov. 1929).
Lawand, T.A., *Saline Water Demineralization with Solar Energy: A Review*, Internal Report No. I.3. of Brace Research Institute, Canada, p.18 (May 1960).
Lawand, T.A., *A Description of the Construction of Solar Demineralization Still No. 1*, Technical Report No. T-1, Brace Research Inst., McGill University, 21 pages (February 5, 1962).

Lawand, T.A., Producing a wettable surface on tedlar film, Memo to File 41/04, Brace Research Institute, McGill University (August 4, 1965).
Lawand, T.A., The preparation of the basin for a solar still, Memo to File 41/04, Brace Research Institute, McGill University (August 6, 1965).
Lawand, T.A., Notes on the construction and operation of large plastic solar still, No. SSP-1, Internal Report No. 1-29, Brace Research Institute, McGill University, 30 pages (Sept. 1965).
Lawand, T.A., *The Economics of Wind Powered Desalination Systems*, Technical Report No. T-36, Brace Research Institute, McGill University, 56 pages (September 1967).
Lawand, T.A., *Instructions for Constructing a Simple 8 Sq. ft. Solar Still for Domestic Use and Gas Station*, Technical Report No. T-17, Brace Research Institute, McGill University, 6 pages (Revised Sept. 1967).
Lawand, T.A. and Thierstein, G.E., *The Economic Consequences of the Installation of a Water Supply System in a Small Underdeveloped Community*, Miscellaneous Report No. M-22, Brace Research Institute, McGill University, 22 pages (April 1968).
Lawand, T.A., Summary of meeting with W.R. Read of CSIRO, Australia, Internal Memo No. I-41 (File 60/08), Brace Research Institute, McGill University, 6 pages (April 29-30, 1968).
Lawand, T.A., Summary of meeting with Church World Service, Internal Memo to file 60/08, Brace Research Institute, McGill University, 3 pages (May 1, 1968).
Lawand, Thomas Anthony, *Engineering and Economic Evaluation of Solar Distillation for Small Communities*, Technical Report No. MT-6, Brace Research Institute, McGill University, 262 pages (August 1968). (Presented at Solar Energy Society meeting, Palo Alto, California (Oct. 1968).
Lawand, T.A., *Proposal for the Installation of a Solar Distillation Plant on De de la Gonave, Haiti*, Internal Report No. I-46, Brace Research Institute, McGill University, 7 pages (Oct. 1968).
Lawand, T.A. and Alward, R., *Plans for a Glass and Concrete Solar Still*, Technical Report No. T-58, Brace Research Institute, McGill University, 21 pages (Dec. 1968).
Lawand, T.A., *Solar Distillation for Small Communities*, Brace Research Institute, Quebec Technical Report No. MT-6 (1968).
Lawand, T.A., The technical evaluation of a large-scale solar distillation plant, Paper presented at annual ASME meeting, Los Angeles, California, Paper No. 69-WA/SOL-8, 30 pages (Nov. 1969).
Lawand, T.A. and Boutiere, H., Solar distillation: Its application in arid zones in water supply and agricultural production, The First World Symposium on Arid Zones, Mexico (Nov. 1970).
Lawand, T.A. and Alward, R., *Plans for a Glass and Concrete Solar Still* (Technical Report), Brace Research Institute, December 1968, Revised October 1972.
Lawand, T.A., Systems of solar distillation, Brace Research Institute, (Paper presented at conference), September, 1975.
Lawand, T.A., Solar stills in the West Indies, *Proc. International Solar Energy Congress*, Melbourne, Australia (Extracts from the Conference Bulletin No. 2) (1979).
Lejeune, G. and Savornin, I., Solar distillation of water in Algeria (in French), *Journal de Physique, Paris*, $\underline{15}$, 525 (June 1954).
Lejeune, G., Avoiding crystallization in solar stills, *Trans. Conference on the Use of Solar Energy*, Tucson, Arizona, $\underline{3}$, 142-144 (1955).
Lejeune, G., Formation of scale, and incrustation in solar stills (in French), U.N. Conference on New Sources of Energy, Paper 35/S/89, Rome, 7 pages (August 1961).

Leslie Salt Co., *A Study of the Technical and Economic Feasibility of Recovering Fresh Water from Sea Water in Conjunction with Solar Salt Producing Operations*, Report to OSW, 67 pages (Dec. 15, 1962).

Levine, Sumner N., *Selected Papers on Desalination and Ocean Technology*, Dover Publications, NY (1968).

Linermore, S., Report of preliminary tests of sea water still, Staten Island, New York, *Proc. World Symposium on Applied Solar Energy*, Phoenix, Arizona (1955).

Liu, B.Y.H. and Jordan, R.C., The interrelationship and characteristic distribution of direct diffuse and total solar radiation, *Solar Energy*, 4(3), 1-19 (July 1980).

Lof, G.O.G., *Demineralization of Saline Water with Solar Energy*, OSW Report No. 4; PB 161379, 80 pages (August 1954).

Lof, G.O.G., *Solar Distillation of Sea Water in the Virgin Islands*, OSW Report No. 5, PB 161380, 39 pages (Feb. 1955).

Lof, G.O.G., Design and cost factors of large basin-type solar stills, *Proc. Symposium on Saline Water Conversion*, National Academy of Sciences, Nation Research Council, Washington, D.C., Publication No. 568, 157-174 (Nov. 4-6, 1957).

Lof, G.O.G., *Solar Distillation Pilot Plant - Design Modification in Deep-basin Still*, Report to OSW (Apr. 1, 1958).

Lof, G.O.G., *Design and Evaluation of Deep-basin, Direct Solar Heated Distiller for Demineralization of Saline Water*, Final Report to OSW, 60 pages (June, 1959).

Lof, G.O.G., Design and operating principles in solar distillation basins, *Saline Water Conversion, Advances in Chemistry*, Serial No. 27, American Chemical Society, Washington, D.C., 156-165 (1960).

Lof, George O.G., Review of desalination processes - solar distillation, *J. Amer. Water Works Assn.*, 52(5), 578-584 (May 1960).

Lof, G.O.G., Solar distillation of saline water, *Southwest Waterworks Journal, Texas*, 1-4 (July 1960).

Lof, G.O.G., Solar distillation, *Journal of American Waterworks Association*, 52(8), 1-7 (1960).

Lof, G.O.G., Fundamental problems in solar distillation, *Proc. Symposium on Research Frontiers in Solar Energy Utilization*, National Academy of Sciences, Washington, D.C., 47, 1279-1289 (April 1961) and Solar Energy (Special issue), 5, 35-46 (Sept. 1961).

Lof, George O.G., Application of theoretical principles in improving the performance of basin-type solar distillers, U.N. Conference on New Sources of Energy, Paper 35/S/77, Rome, 17 pages (August 1961).

Lof, G.O.G., Eibling, J.A. and Bloemer, J.W., Energy balances in solar distillers, *AIChE Journal*, 7(4), 641-649 (December 1961).

Lof, George O.G., Final report on consultant contract of Feb. 1, 1961, including progress report for Jan-March, 1962, Unpublished report to OSW, Washington, D.C. (1962).

Lof, G.O.G., *Demineralization of Saline Water* (Russian translation), IL, Moscow, 1963.

Lof, George O.G., Solar distillation, Chapter 5 in *Principles of Desalination*, edited by K.S. Spiegler, Academic Press, New York, 151-198 (1966).

Lof, George O.G., Duffie, John A. and Smith, Clayton O., World distribution of solar radiation, *Solar Energy*, 10(1), 27-37 (1966).

Lof, George O.G., Duffie, John A. and Smith, Clayton O., *World Distribution of Solar Radiation*, Report No. 21, Engineering Experiment Station, University of Wisconsin, 72 pages (July 1966).

Lof, George O.G., Letter to Editor on solar stills, *Sun at Work*, 18-19 (Fourth Quarter, 1966).

Lof, G.O.G., Comparison of design and performance of large solar distillation plants of the world, oral presentation at ASME meeting, New York City (Nov. 28, 1966).

Selected Bibliography

Lof, George O.G., Desalting of sea water with solar energy, Paper presented at IECEC, Boulder, Colorado, 18 pages (Aug. 1968).

Lof, George, O.G., *Technical Cooperation in the Solar Distillation Development Program of Spain*, OSW Report No. 397, 25 pages (Sept. 1968). (Published in part by Distillation Digest, $\underline{2}$(3), 39-49 (Fall, 1968).

Lof, George O.G., The economics of water desalting with solar energy, Chapter prepared for U.N. Solar Distillation Panel Meeting, 41 pages (Oct. 14-18, 1968).

Lof, George O.G., Experience in community supply of solar distilled water, Paper at Solar Energy Meeting in Palo Alto, California (Oct. 1968).

Lof, G.O.G., Close, D.J. and Duffie, J.A., A philosophy for solar energy development (presented at 1966 Solar Energy Society Conference), *Solar Energy*, $\underline{12}$(2), 243-250 (Dec. 1968).

Lof, George O.G., Letter to Editor (Regarding earlier paper by Morse and Read, *Solar Energy*, $\underline{12}$, 1968), *Solar Energy Journal*, $\underline{12}$(4) (1969).

MacLeod, L.H. and McCracken, H., *Performance of Greenhouse Solar Stills*, Sea Water Conversion Program, University of California, Series 75, Issue 26, Contribution 46, 57 pages (Nov. 30, 1961).

MacLeod, L.H. and McCracken, H.W., Correlation of the effects of temperature, geometry and heat capacity on the performance of a single-effect solar distiller, Saline Water Conversion Research, Contribution 47, University of California, 26 pages (Nov. 30, 1961).

MacLeod, L.H., Tleimat, B.W. and Howe, E.D., Performance of single-effect solar stills, Paper presented at Solar Energy Symposium, University of Florida, Gainesville, Florida (April 1963).

McCracken, H., Performance of natural conversion stills, Sea Water Conversion Laboratory, University of California, Internal Report, 10 pages (1960).

McCracken, Horace, The inclined tray solar still, *Sun at Work*, 6-9 (Third Quarter, 1963).

McCracken, H.W., Solar still pans: The search for inexpensive and reliable materials, *Solar Energy*, $\underline{9}$(4), 201-207 (1965).

McCracken, H., Description of sunwater company solar stills, San Diego, California, 1968.

Maheswari, G.K. and Sahgal, P.N., Development of a high efficiency solar water still, *Proc. ISES*, New Delhi, India, $\underline{3}$, Series 51, Abstract No. 8, p.1441 (1978).

Malik, M.A.S. and Tran, Van Vi, Digital simulation of nocturnal production of a solar still, ASME Winter meeting, New York, Paper No. 70-WA/Sol-6 (1970).

Malik, M.A.S. and Tran, V.V., A simplified mathematical model for predicting the nocturnal output of a solar still, *Solar Energy*, $\underline{14}$, 371 (1973).

Malik, M.A.S. and Puri, V.M., Nocturnal production in a double stage solar still, Paper presented at International Desalination Conference, Spain (1978).

Malik, M.A.S., Puri, V.M. and Aburshaid, H., Use of double stage solar stills for nocturnal production, *Proc. 6th International Symposium, Fresh Water from the Sea*, $\underline{2}$, 367 (1978).

Manchester, H., They're turning salt water into sweet, *Reader's Digest*, 149 (May 1965).

Mansfield, W.W., The influence of mono layer on evaporation from water storages. I. Evaporation and seepage from water storage, *Australian J. Applied Science*, $\underline{10}$(1), 65 (1959).

Marshak, I., Distillation of sea-water, *Power Engg.*, $\underline{55}$, 67 (Feb. 1951).

Artens, C.P., Theoretical determination of flux entering solar stills, *Solar Energy*, 10(2), 77-80 (Apr.-June 1966).

Masson, Henri, Solar stills, *South Pacific Commission, Quarterly Bulletin*, 33-37 (January 1957).

Masson, H., Quantitative estimation of solar radiation, *Solar Energy*, $\underline{10}$(3), 119-124 (1966).

Masson, H., Personal communication (Apr. 22, 1969).
Mathur, K.N., Solar energy - Its measurement and utilization, 37th Annual Session, National Academy of Sciences, Ahmedabad, India (Feb. 1-4, 1968).
Maurian, Ch. and Brazier, C. (Title unknown), *Recherches et Inventions*, No. 654 (1927).
Maurian, Ch. and Brazier, C. (Title unknown), *Recherches et Inventions*, No. 173 (1929).
Mehta, M.H., Development of air-recycled H-D technique of water desalination, Presented at the Annual Meeting of Indian Institute of Chemical Engineers, Hyderabad, Dec. (1970).
Meigs, Peveril, Coastal deserts: Prime customers of desalination, *Proc. First Int'l. Symp. on Water Desalination*, Washington, D.C. (Oct. 1965), 3, 721-736 (1967).
Menguy, G., Chassagne, G., Sfeir, A. and Saab, J., Experimental study and optimization of a solar still, *Revue Internationale d'Heliotechnique*, p.46 (1976) (in French).
Menguy, G., Benoit, M., Louat, R., Makki, A. and Schwartz, New solar still design and experimentation (The wiping spherical still), private communication, Group d'Etudes Thermiques et Solaires, Universite Claude Bernard 43 Bd du 11 Novembre 1918, 69622 Villeurbanne - Cedex France (1980).
Menjarrez, R. and Galvan, M., Solar multistage flash evaporation (SMSF) as a solar energy application on desalination processes - Description of one demonstration project, *Proc. Int. Cong. on Desalination and Water Reuse*, 1979, Published by Elsevier Scientific Publications, Amsterdam, Netherland and NY, 31, 545 (1979).
Morris, L.G., Trickett, E.S., Vanstone, F.H. and Wells, D.A., The limitation of maximum temperature in a glasshouse by the use of water film on the roof, *J. Ag. Engng Res.*, 3(2), 129 (1968).
Morse, R.N. and Read, W.R.W., A solar still for Australian conditions, Paper at Conf. on Power Production and Energy Conversion, The Instn. of Engrs., Sydney, Australia (1966).
Morse, R.N. and Read, W.R.W., The engineering development of a solar still on a rational basis, Paper presented at Solar Energy Conference, Boston (March, 1966).
Morse, R.N., The potential for solar distillation in Australia, *Proc. Int'l. Conference on Water for Peace*, Washington, D.C., 122 (May 23-31, 1967).
Morse, R.N., The construction and installation of solar stills in Australia, Second European Symposium on Fresh Water from the Sea, Athens, Greece, Preprints of Papers, 7, Paper No. 101, 8 pages (May 1967).
Morse, R.N. and Read, W.R.W., The development of a solar still for Australian conditions, *Trans. Institution of Engineers, Mechanical and Chemical Engineering*, Australia, MC3(1), 77-80 (May 1967).
Morse, R.N. and Read, W.R.W., A rational basis for the engineering development of a solar still, *Solar Energy*, 12(1), 5-17 (1968).
Morse, R.N., The construction and installation of solar stills in Australia, *Desalination*, 5(1), 82-89 (1968).
Morse, R.N., Solar distillation in Australia, *Civil Engr.*, 38, 39-41 (Aug. 1968).
Morse, R.N., Read, W.R.W. and Trayford, R.S., Operating experiences with solar stills for water supply in Australia, Paper presented at Solar Energy Society Annual Meeting, Palo Alto, California, 6 pages (Oct. 21-23, 1968).
Morse, R.N., The theory of solar still operation, Chapter prepared for U.N. Solar Distillation Panel Meeting, 12 pages (Oct. 14-18, 1968).
Morse, R.N., Read, W.R.W. and Trayford, R.S., Operating experiences with solar stills for water supply in Australia, *Solar Energy*, 13, 99 (1970).
Morse, R.N., Read, W.R.W. and Trayford, R.S., Solar still operation in Australia, International SES Conference, Melbourne, Paper 5/34 (1970).

Selected Bibliography

Morse, R.N., Read, W.R.W. and Trayford, R.S., Solar still operation, *Proc. International Solar Energy Congress*, Melbourne, Australia (Extracts from the Conference Bulletin No. 2) (1979).

Mostag, M.M., Double exposure solar still, Paper presented at NSF International Symposium - Workshop on Solar Energy, Cairo, Egypt, June 16-22 (1978).

Mouchot, A., *The Solar Heat and its Industrial Applications* (in French), Gauthier-Villars, Paris, 238 pages (1969).

Moustafa, S.M.A. and Brusewitz, G.H., Direct use of solar energy for water desalination, *Solar Energy*, 22, 141 (1979).

Muecke, Madeline M., Taming the energy of the Sun, *Product Engineering*, 40(14), 130 (July 14, 1969).

Murphy, G.W., The minimum energy requirements for sea water conversion process, Research and Development Progress Report No. 9, U.S. Department of the Interior, Office of Saline Water (1956).

Muthuveerappan, V.R. and Kamaraj, G., Mini solar still for rural applications, *Proc. ISES*, New Delhi, India, 3, Series 51, Abstract No. 247, 1535 (1978).

National Academy of Sciences - National Research Council, Saline water conversion, (Sect. 3 - solar distillation), *Proc. Symposium*, Publication 568, 117-176 (Nov. 4-6, 1957).

National Academy of Sciences, *Symposium on Research Frontiers in Solar Energy Utilization*, 47(8) (Aug. 1961).

National Aeronautics and Space Administration, Manned Spacecraft Center, *Emergency Solar Still Desalts Seawater*, Tech. Brief, 65-10214 (July, 1965).

Natu., G.L., Goghari, H.D., and Gomkale, S.D., Solar distillation plant at Awania, Gujarat, India, *Desalination*, 31, 435 (1979).

Nayak, J.K., Transient thermal processes in solar collector/storage and distillation systems, Ph.D. Thesis, Department of Physics, I.I.T. Delhi, New Delhi, India (1980).

Nayak, J.K., Transient theory of single as well as double stage solar stills, 3rd Miami Int. Conf. on Alternative Energy Sources (Dec. 1980).

Nayak, J.K., Singh, U. and Tiwari, G.N., Thermal performance of a solar still, *J. of Energy* (1980) (in press).

Nayak, J.K., Tiwari, G.N. and Sodha, M.S., Periodic theory of solar still, *Int. J. of Energy Research*, 4, 41 (1980).

Nebbia, G., Research on solar distilllation (in Italian). *Della Camera di Commercio Industria e Agricoltura di Bari*, 37 (11/12), 7 pages (Nov.-Dec. 1953).

Nebbia, Giorgio, On the utilization of solar energy (in Italian), *Geofis. e Meteor.*, 1(6), 100-102 (Nov.-Dec. 1953).

Nebbia, Giorgio, Solar heat for the production of potable water in a solar still (in Italian), *Le Vie J'Italia*, 3, 400 (Mar. 1954).

Nebbia, G., Some new studies on solar distillation (in Italian), *La Chimica e l'Industria, Milani*, 36, 20-27 (1954).

Nebbia, Giorgio, Utilization of solar energy, (in Italian), *Geofis. e Meteor.*, 2, 50-51 (1954).

Nebbia, Giorgio, A solar still (in Italian), *COELVM*, 22, 4 pages (Sept.-Oct. 1954).

Nebbia, G., A new type of solar still (in Italian), *La Ricerca Scientifica, Roma*, 25(6), 1443-1446 (June 1955).

Nebbia, G., An experiment with a plastic tabular solar still, *Proc. Symp. on Saline Water Conversion*, National Academy of Sciences, National Research Council, Washington, D.C., Publication No. 568, 175-176 (Nov. 4-6, 1957).

Nebbia, G., Experiments with a tubular solar distiller made of plastic material (in Italian), *Macchine e Motori Agricoli*, 16(7), 83-87 (July 1958).

Nebbia, Giorgio, An experiment with a vertical solar still, Paper presented at the UNESCO-Iran Symposium on Salinity Problems in Arid Zones, Tehran, 3 pages (Oct. 11-15, 1958).

Nebbia, G., Transforming brackish water into soft water (in Italian), *Bollettino Scientifico Della Facolta Di Chimica Industriale, Bologna,* 16, 44-63 (1958).

Nebbia, Giorgio and Pizzoli, Elsa M., Research on a vertical solar still (in Italian), *La Ricerca Scientifica,* 29(9), 1941-1945 (Sept. 1959).

Nebbia, Giorgio, Present status and future of the solar stills, U.N. Conference on New Sources of Energy, Paper 35/S/113, Rome, 12 pages (August 1961).

Nebbia, G., The experimental work on the solar stills in the University of Bari, *Proc. International Seminar on Solar and Aeolian Energy,* Sounion, Greece, September 1961, Plenum Press, New York, 19-27 (1964).

Nebbia, G., Researches in the University of Bari (Italy), *C.O.M.P.L.E.S.,* Bull. No. 5, 26 (1963).

Nebbia, Giorgio, *The Demineralization of Saline Water with Solar Energy,* Roma, Consiglio Nazionale Delle Ricerche, 145-149 (1963).

Nebbia, G. and Menozzi, G., A short history of water desalination, Acque Dolce Dal Mare, II^a Inchiesta Internazionale (*Proc. International Symposium,* Milano, Apr. 1966), 129-172 (1967.

Nebbia, Giorgio, Present status of desalination techniques and prospective of use in Sicily (in Italian), *Industria Grafica Nazionale, Palermo,* 31-75 (Sept. 10-12, 1966).

New York University, *Research on Methods for Solar Distillation,* United States Department of the Interior, Office of Saline Water, R & D Report No. 13, pp. 1-65 (1956).

Niaz, R.H., Gwadar solar desalination project, *The Nucleus (Journal of the Pakistan AEC),* 5 (1 and 2), 38-41 (Jan.-June 1968).

Norov, E. Zh., Experimental investigations of solar stills with various film surfaces, *Applied Solar Energy (Geliotekhnika),* 13(1), 61 (1977).

Office of Saline Water (1956), *A Standard Procedure for Estimating Costs of Saline Water Conversion,* Office of Saline Water, Washington.

Ottra, F., *Saline Water Conversion and Its Stage of Development in Spain,* Publications of J.E.N., Madrid (1972).

O'Shaughnessy, F., *Desalting Plants Inventory Report No. 3,* O.S.W., U.S. Department of the Interior, Washington, D.C. (1972).

Othmer, Donald F., Fresh water, energy, and food from the sea and the sun, *Desalination,* 17(2), 193 (1975).

Oztoker, U. and Selcuk, M.K., *Theoretical Analysis of a System Containing a Solar Still with a Controlled Environment Greenhouse,* ASME Paper No. 71/WA/SOL/9, 28 (Nov. 1971).

Parmelee, G.V., The transmission of solar radiation through flat glass under summer conditions, *Heating, Piping and Air-Conditioning,* ASHVE J. Section, 17(10), 562 (October/November, 1945).

Parmelee, G.V. and Aubele, W.W., Radiant energy emission of atmosphere and ground, *Heating, Piping and Air-Conditioning,* ASHVE J. Section, 123 (Nov. 1951).

Parmelee, G.V., Irradiation of vertical and horizontal surfaces by diffuse solar radiation from cloudless skies, *Heating, Piping and Air-Conditioning,* ASHVE J. Section, 1-7 (June 1954).

Petersen, George G., On the footsteps of Aldexander von Humboldt, his geological and geophysical observations in Peru, (in French), *Publications de l'Institut de Geographic de l'Universite de San Marcos, Lima,* 118 (1960).

Petersen, G., Fries, S., Mohn, J. and Mueller, A., Wind and solar-powered reverse osmosis desalination units - description of two demonstration projects, *Proc. Int. Cong. on Desalination and Water Reuse,* 1979. Published by Elsevier Scientific Publications, Amsterdam, Netherland and NY, 31, 501 (1979).

Poullain; Ginestous, Pasteur; Pouget, Meeting relating to the distillation of water with solar energy (in French), *Recherches et Inventions,* 8, 205-215 (June 1927).

Powell, S.T., Saline water conversion, *Water and Sewage Works*, 106(2), 84-86 (Feb. 1959).
Pressel, F., Methods for desalting sea water and brackish water (in German), *VDI Zeit.*, 98(1), 9-13 (Jan. 1, 1956).
Pretoria, P.W.D., Drawings for solar still, C.S.I.R., No. CR 789 (1954).
Proceedings, Organisation Europeene de cooperation economique, Alger (May 3-5, 1955).
Proceedings, World Symposium on Applied Solar Energy, Phoenix, Arizona, Stanford Research Inst., Menlo Park, California (Nov. 1955).
Proceedings, Symposium on Saline Water Conversion, Washington, D.C. (Nov. 1957), National Academy of Sciences National Research Council, Publication 568 (1958).
Proceedings, Desalination Research Conference, Woods Hole, Massachusetts (June-July 1961), U.S. Dept. of the Interior, OSW, Publication 942 (1963).
Proceedings, United Nations Conference on New Sources of Energy, Rome (Aug. 21-31, 1961), U.N. Document E/CONF. 35/2, 6, 139 pages (1964).
Proceedings, International Seminar on Solar and Aeolian Energy, Sounion, Greece (Sept. 1961), Plenum Press, New York (1964).
Proceedings, First International Symposium on Water Desalination, Washington, D.C. (Oct. 1965), 3 volumes, U.S. Dept. of the Interior, OSW.
Proctor, D., The use of waste heat in a solar still, *Solar Energy*, 14(4), 433-449 (1973).
Proctor, D., Supplementary heat in solar stills, *Proc. International Solar Energy Congress*, Melbourne, Australia (Extracts from the Conference Bulletin No. 27 (1979).
Public Works Dept., Pretoria, Union of South Africa, Drawings for solar still, C.S.I.R. Drawing No. CR 789 (1954).
Qasim, S.R., Treatment of domestic sewage by using solar distillation and plant culture, *Journal of Environmental Science and Health*, 13(8), 615 (1978).
Radway, Wilson, Personal communication (Jan. 20, 1969).
Rajan, S.T. and Sivanandan, C., A new method for solar desalination of sea water, *Proc. ISES*, New Delhi, India, 3, Series 51, Abstract No. 198, 1519 (1978).
Rajvanshi, Anil, K. and Hseieh, C.K., Effect of dye on solar distillation: Analysis and experimental evaluation, *Proc. International Congress of ISES*, at Georgia, Atlanta, p.20, 327 (1979).
Ramsay, N., Suggestions for solar distillation of saline water, Nairobi, Kenya, unpublished, 10 pages (1954).
Raseman, Chad J., Personal communication (Aug. 8, 1969).
Raymond, C.H., Production of water from the atmosphere, *Solar Energy*, 10(1), 41 (1966).
Read, W.R.W., A solar still for water desalination, Australian C.S.I.R.O., Division of Mechanical Engineering, Report No. 9, 23 pages (Sept. 1965).
Read, W., Personal communication (1968).
Read, W.R.W. and Trayford, R.S., Solar distillation, *Solar Energy Progress in Australia and New Zealand*, Publication of the Australian and New Zealand Section of the Solar Energy Society, No. 8, 19-21 (July 1969).
Read, W.R.W., Recent developments and future trends in solar distillation, 1970 ISES Conference, Melbourne, Australia, Paper No. 5/52 (1970).
Read, W.R.W., The economics of solar distillation, *Proc. International Solar Energy Congress*, Melbourne, Australia (Extracts from the Conference Bulletin No. 2) (1979).
Read, W.R.W., Development and trends in solar distillation, *Proc. International Solar Energy Congress*, Melbourne, Australia (Extracts from the Conference Bulletin No. 2) (1979).
Reichle, C. and Keller, H., Production of fresh water from sea water and similar saline waters in hot countries, *Kl. Mitt. Ver. Wassen-Boden-U. Lufthyg.*, 19, 1 (1943). (Abst. in *J. Am. Water Works Assn.*, 43(5), 56 (1951)).

Rheinlaender, J., Analogy between heat and mass transfer in calculating solar stills, *Proc. Int. Symp. on Fresh Water from the Sea*, 5th Event of the Eur. Fed. of Chem. Engg., 162nd, Alghero, Italy $\underline{2}$, 467-476 (1976).

Richard, J., Solar distillation of saline waters (in French), *Recherches et Inventions*, $\underline{8}$, 474-475 (July 1927).

Richard, J., Comments on a simple process to extract soft water from sea water, from salt and polluted waters, from any aqueous substances and even from the air, by means of the solar heat (in French), *Bulletin de l'Institut Oceanographique*, No. 535, Monaco, 27 pages (April 1929).

Richard, J., Extraction of fresh water from saline waters and from any aqueous substances by means of solar heat (in French), *Recherches et Inventions*, No. 173, 18-19 (1929).

Richard, J., (Title unknown), *La Nature*, No. 2804 (1929).

Rocca, A. La, Desalination of salt water by solar energy means, US Patent 4, 135, 985 (Jan. 23, 1979). File date: May 31, 1976.

Sakr, I.A., Portable solar water distiller, Second European Symposium on Fresh Water from the Sea, Athens, Greece, Paper No. 142, 5 pages (May 1967).

Sakr, I.A., Empirical formula for the expected fresh water production by solar energy, *COMPLES? Marseilles*, Bull. No. 12, 56 (July 1967).

Sakr, I.A., Economic investigation of solar water distillation in Egypt, *Proc. Int. Symp. on Fresh Water from the Sea*, $\underline{2}$, 477 (1976).

Salam, Ehab and Daniels, F., Solar distillation of salt water in plastic tubes using a flowing air stream, *Solar Energy*, $\underline{3}$(1), 19-22 (Jan. 1959).

Salieva, R. B. and Baitulakova, N.D., Method of calculating the optimum parameters of a still system, *Applied Solar Energy (Geliotekhnika)*, $\underline{12}$(2), 18 (1976).

Samuel, J., Desalination: Water for urban and industrial growth, *Energy Business*, $\underline{2}$(1), 16-19 (1980).

Satcunanathan, S. and Hansen, H.P., An investigation of some of the parameters involved in solar distillation, *Solar Energy*, $\underline{14}$, 353 (1973).

Satcunanathan, S., Remarks on 'cascade' solar still for distilled water production, *Sol. Energy*, $\underline{17}$(1), 81 (1975).

Savornin, J., Efficiency of various types of soalr stills, *Trans. Conference on Use of Solar Energy*, Tucson, Arizona, $\underline{3}$, 134-137 (1955).

Savornin, J. and Lejeune, G., Attempts to perfect a solar distillation apparatus (in French), *Comptes Rendus de l'Academie des Sciences*, Paris, $\underline{243}$(1), 32-34 (July 2, 1956).

Sayigh, A.A.M. and El-Salm, E.M.A., Optimum design of a single slope solar still in Riyadh, Saudi Arabia, *Revue d'Heliotechnique*, $\underline{1}$, 40 (1977).

Seiitkurbanov, S. and Rabinovich, L.I., Effect of closed-loop forced circulation of vapor-air mixture in a nozzle type still, *Applied Solar Energy*, $\underline{14}$(2), 36 (1978).

Selcuk, M.K., *Design and Performance, Evaluation of a Multiple Effect Tilted Solar Distillation Unit*, Brace Experiment Station of McGill University, St. James, Barbados, pp.1-29 (1963).

Selcuk, M. Kudret, Design and performance evaluation of a multiple-effect, tilted solar distillation unit, *Solar Energy*, $\underline{8}$(1), 23-30 (1964).

Selcuk, K., The role of solar distillation in the saline water conversion processes, *COMPLES, Marseilles*, Bulletin No. 8, 70-78 (May 1965).

Selcuk, M.K., The effect of solar radiation on the energy balance of a controlled environment greenhouse, ASME Annual Winter Meeting, New York, Paper No. 70/WA/5063 (1970).

Selcuk, M.K., Analysis, design and performance evaluation of controlled environment greenhouse, *Trans. ASHRAE*, No. 2172 (1971).

Selcuk, M.K., Greenhouses and solar stills combines, *COMPLES Conference*, Athens, Bull. No. 22 (1971).

Selcuk, M.K. and Tran, V.V., An overview of solar still greenhouse performance and optimal design studies, *Heliotechnique and Development*, edited by M.A. Kettani and J.E. Soussou, Vol. II, 349, DAA (1976).

Sharafi, A.Sh., Continuous solar still, *Applied Solar Energy* (trans. of *Geliotekhnika*), 1(2), 37-39 (March-April 1965).

Sharma, M.R. and Pal, R.F., Total, direct and diffuse radiation in the tropics, *Solar Energy*, 9(4), 183-192 (March 1965).

Singh, Daljit and Gupta, Y.P., Thermodynamics of solar flash evaporation of sea water, *Proc. National Solar Energy Convention*, Calcutta, India, 171 (1976).

Singh, Daljit and Gupta, Y.P., Thermoeconomics of solar flash evaporation of sea water, *Proc. ISES*, New Delhi, India, 3, Series 51, Abstract No. 30, 1453 (1978).

Sobornin, J., Efficiency of various types of solar stills, *Trans. Conf. on the Use of Solar Energy, The Scientific Basis*, Tucson, Arizona, 3, 134 (1955).

Sodha, M.S., Kumar, A., Singh, U. and Tiwari, G.N., Further studies on double basin solar still, *Int. J. Energy Research* (1981).

Sodha, M.S., Kumar, A., Singh, Usha and Tiwari, G.N., Transient analysis of solar still, *Energy Conversion*, 20(3), 191 (1980).

Sodha, M.S., Kumar, A., Srivastava, A. and Tiwari, G.N., Thermal performance of 'Still-on-roof' system, *Energy Conversion*, 20(3), 181 (1980).

Sodha, M.S., Kumar, A., Tiwari, G.N. and Tyagi, R.C., Simple multiple-wick solar still: Analysis and performance, *Solar Energy*, 26, 127 (1981).

Sodha, M.S., Kumar, Ashwini, Tiwari, G.N. and Pandey, G.C., Effect of dye on the still performance, *Applied Energy*, 7, 147 (1980).

Sodha, M.S., Nayak, J.K., Tiwari, G.N. and Kumar Ashwini, Double basin solar still, *Energy Conversion*, 20(1), 23 (1980).

Sodha, M.S., Singh, Usha, Kumar, A. and Tiwari, G.N., Enhancement of output in a double basin solar still, National Solar Energy Convention, Annamalai, TN, India (1980).

Sodha, M.S., Kumar, A. and Tiwari, G.N., Utilization of waste hot water for distillation, *Desalination* (1981) 37 (In Press).

Solar Sunstill, Inc., *Assembly Instructions for Model 64*, Setauket, Long Island, New York (1969).

Solar Sunstill, Inc., *The Solar Sunstill*, Setauket, Long Island, New York, 7 pages (July 30, 1969).

Soliman, S.H., Water distillation by solar energy, Doctor Thesis, Faculty of Engineering, Keio University, Tokyo (1967).

Soliman, S.H. and Kobayashi, M., Water distillation by solar energy, Paper presented at 1970 International Solar Energy Society Conference, Melbourne, Australia.

Solinan, S.H., Effect of wind on solar distillation, *Solar Energy*, 13, 403 (1972).

Soliman, S.H., Solar still coupled with a solar water heater, *Revue Internationale d'Heliotechnique*, 1er Semestre, 43 (1979).

Soliman, S.H. and Kobayashi, M., Water distillation by solar energy, *Proc. International Solar Energy Congress*, Melbourne, Australia (Extracts from the Conference Bulletin No. 2) (1979).

Sommerfeld, J.V., Informe sobre trabajos de investigacion en destiladores solares que se realizan en la U.S.M., *Scientia*, 139, Valparaiso, Chile (1970).

Spiegler, K.S., *Salt Water Purification*, John Wiley and Sons, New York (1962).

Spiegler, K.S. (editor), *Principles of Desalination*, Academic Press, N.Y. (1966), Chapter 5, Solar Desalination (George O.G. Lof), 151-198.

Stanley, W.E., *Economic Considerations of Fresh Water by Solar Distillation*, Massachusetts Inst. Technology - Solar Energy Research Project (2743), 74 pages (1957).

Stober, W.J. and Falk, L.H., A benefit-cost analysis of local water supply, *Land Economics*, 43(3), 328-335 (August 1967).
Strobel, J.J., Summary of solar distillation processes, *Proc. Symposium on Saline Water Convention*, National Academy of Sciences, Nat'l. Research Council, Publ. No. 568, 117-122 (Nov. 4-6, 1957).
Strobel, Joseph J., Sieveka, Ernest, H., Sandell, Dewey J., Jr., Bradt, David M. and Lof, George O.G., Review of desalinization processes (panel discussion), *Journal American Water Works Association*, 52(5), 32 pages (May 1960).
Strobel, J.J., Saline water conversion today, *Chem. Eng. Progress*, 57(1), 37-41 (Jan. 1961).
Strobel, J.J., Developments in solar distillation, U.N. Conf. on New Sources of Energy, Paper No. 35/S/85, Rome, 45 pages (1961).
Suarez, Jorge and Pliego, Jose M., Desalination in Spain, *Proc. First Internat'l. Symp. on Water Desaliantion*, Washington, D.C. (Oct. 3-9, 1965), 2, 797-811 (1967).
Swaiden, B.E., Solar desalination, *Alternative Energy Sources*, Miami International Conference, Int. Compend., Miami Beach, Florida, 1977. Published by Hemisphere Publications Corporation, Washington, D.C., 2, 951 (1978).
Swaiden, Brian E., High performance solar still, US Patent Application 003 181 (CAD-D 005 723/2). Filed data Jan. 15, 1979.
Szulmayer, W., Solar stills with low thermal inertia, *Solar Energy*, 14(4), 415 (1973).
Tabor, H., The status of solar energy, *Sun at Work*, 10(2), 3-7 (1965).
Talbert, S.G., Eibling, J.A. and Lof, G.O.G., *Manual on Solar Distillation of Saline Water*, R & D Progress Report No. 546, U.S. Department of the Interior (1970).
Tarnekar, M.G. and Zadgaonkar, A.S., Solar-MHD generator, *Proc. ISES*, New Delhi, India, 3, Series 51, Abstract No. 183, 1512 (1978).
Taylor, Floyd B., Desalting water by solar distillation, *Plumbing Engineering*, 5(3), 34 (1977).
Teagan, W.P. and Cunningham, D.R., Use of solar energy in conventional desalination processes, *Proc. Int. Conference on Solar Building Technology*, London, U.K., 2, 603 (July 1977).
Tekuchev, A. (Title unknown), *Trudy Usbekskogo Universiteta*, No. 59, Samarkand (1935).
Tekuchev, A.N., Physical basis for the construction and calculation of a solar still with a finned surface, *Trans. Uzbekistan State Univ., Samarkand*, 2 (1938).
Tekuchev, A.A., Physical foundation for the construction and calculation of solar stills with a ribbed surface, *Trudy Uzbekskogo, gosuniversiteta*, No. 59, Samarkand (1955).
Telkes, Maria, *Distilling Water with Solar Energy*, Report to the Solar Energy Conversion Committee, Massachusetts Institute of Technology (Jan. 1943).
Telkes, Maria, Solar distiller for life rafts, U.S. Office of Science Report No. 5225, P.B. 21120, 24 pages (June 19, 1945).
Telkes, Maria, *Solar Distillation to Produce Fresh Water from Sea Water*, Massachusetts Institute of Technology, Solar Energy Conversion Project, Publication No. 22, 34 pages (Apr. 6, 1951).
Telkes, Maria (Title unknown), Report to the Pacific Science Board, National Research Council (1951).
Telkes, Maria, Fresh Water from sea water by solar distillation, *Industrial and Engineering Chemistry*, 45(5), 1108-1114 (May, 1953).
Telkes, Maria, Solar distillation, *Proc. Symp. on Solar Energy*, Johnson Press, New York (1955).
Telkes, M., Distillation with solar energy, *Proc. World Symposium on Applied Solar Energy*, Phoenix, Arizona, pp.73-79, 1955.

Telkes, Maria, Improved solar stills, *Trans. Conference on the Use of Solar Energy*, Tucson, Arizona, 3, 145-153 (1955).
Telkes, Maria, Distillation with solar energy, in *Solar Energy Research*, ed. by F. Daniels and J.A. Duffie, University of Wisconsin Press, Madison, Wisconsin, 111-118 (1955).
Telkes, Maria, Solar stills, *Proc. World Symposium on Applied Solar Energy*, Phoenix, Arizona, 73-79 (Nov., 1955).
Telkes, Maria, *Research on Methods for Solar Distillation*, OSW Report No. 13, PB 161388, 66 pages (Dec. 1956).
Telkes, M., Solar still theory and new research, *Proc. Symposium on Saline Water Conversion*, National Academy of Sciences, National Research Council, Washington, D.C., Publication No. 568, 137-149 (Nov., 1957).
Telkes, M., *Solar Still Theory and New Research*, NAS-NRO, 568, 137 (1958).
Telkes, M., *Solar Still Construction*, United States Department of the Interior, Office of Saline Water, R & D Report No. 33, pp.1-14 (1959).
Telkes, M., *New and Improved Methods for Lower Cost Solar Distillation*, United States Department of the Interior, Office of Saline Water, R & D Report No. 31, pp.1-38 (1959).
Telkes, M., Private communication (1979).
Telkes, Maria, *New and Improved Methods for Lower Cost Solar Distillation*, OSW Report No. 31, PB 161402, 38 pages (Aug. 1959).
Telkes, Maria, *Solar Still Construction*, OSW Report No. 333, PB 161404, 18 pages (Aug. 1959).
Telkes, M. and Andrassy, S., *Design and Fabrication of Flat, Tilted Solar Stills*, Report to OSW, 55 pages (June 1960).
Telkes, M., Flat, tilted solar stills, *Proc., International Seminar on Solar and Aeolian Energy*, Sounion, Greece, September 1961, Plenum Press, New York, 14-18 (1964).
Telkes, M., Discussion of inflatable plastic stills, *Proc., International Seminar on Solar and Aeolian Energy*, Sounion, Greece, September 1961, Plenum Press, New York, 28-32 (1964).
Thierstein, Gerald E., *Anguilla Solar Still Project*, Internal Report No. I.47, Brace Research Inst., McGill University, 11 pages (Oct. 1968).
Thompson, L. and Hodges, C.N., *Solar Radiation, Water Demand, and Desalinization*, Technical Note, Solar Energy, $\underline{7}$(2), 79-80 (Apr.-June, 1963).
Tinaut, D., Echaniz, G. and Ramos, F., Materials for a solar still greenhouse, *Optica Pura Applicada*, $\underline{11}$, 59 (1978).
Tippetts-Abbett-McCarthy-Straton (Engineers and Architects, NYC), Personal communication (Aug. 8, 1969).
Tleimat, Badawi W. and Howe, Everett D., Nocturnal production of solar distillers, University of California, Sea Water Conversion Lab. Report No. 65-7, Water Resources Center Contribution No. 105, 19 pages (Dec. 1965). (Also in *Solar Energy*, $\underline{10}$(2), 61-66 (Apr.-June 1966).)
Tleimat, B.W. and Howe, E.D., Comparison of plastic and glass condensing covers for solar distillers, *Proc. Solar Energy Society Annual Conference*, Phoenix, Arizona, pp.1-12 (1967).
Tleimat, B.W. and Howe, E.D., Comparison of plastic and glass condensing covers for solar distillers, *Proc., Solar Energy Society Conference*, Phoenix, Arizona (Mar. 1967), *Solar Energy*, $\underline{12}$(3), 293-304 (May 1969).
Tleimat, Badawi W., The effects of construction and maintenance on the performance of solar distillers, paper presented at annual ASME meeting, Los Angeles, California, Paper No. 69-WA/SOL-4, 18 pages (Nov. 1969).
Tleimat, B.W. and Howe, E.D., Solar-assisted distillation of sea water, presented at the COMPLES International Meeting, University of Petroleum and Minerals, Dhahran, Saudi-Arabia, No. 2-6 (1975).
Tleimat, B.W. and Howe, E.D., Solar assisted distillation of sea water, *NWSIA J.*, $\underline{4}$(1), 17 (1977).

Toiliev, K. and Bairamov, R., The influence of heat coefficient on heat regime and still productivity, *Applied Solar Energy (Geliotekhnika)*, 6(4), 65 (1970).

Touchais, M., A new direct exposure solar distiller: The Jean Mary distiller (in French), *COMPLES, Marseilles*, Bulletin No. 12, 64-65 (July 1967).

Touchais, M., Solar marine plants and industrial exploitation of the sea and sea bottom (in French), *COMPLES, Marseilles*, Bulletin No. 13, 20 (Dec. 1967).

Tran, Van-Vi, A study of a solar still coupled with a greenhouse, Ph.D. Thesis, Agr. Eng. Dept., McGill University, Canada (1974).

Tean, V.V., Technical and economical aspects of a solar still coupled with a greenhouse in arid lands of developing countries, *Proc. ISES Congress*, New Delhi, India, 3, 2031 (1978).

Transactions, Conference on the Use of Solar Energy - the Scientific Basis, Tucson, Arizona, Oct. 31-Nov. 1, 1955, University of Arizona Press, 3, Part 2, Solar Distillation, Chapters 8-11, 13-16, pp.99-137, 142-169 (1958).

Trayford, R.S. and Welch, L.W., Dynamic operation of solar stills, *Proc. International Solar Energy Congress*, Melbourne, Australia (Extracts from the Conference Bulletin No.2) (1979).

Trofirnov, K.G., *The Use of Solar Energy in the National Economy* (in Russian), Uzbek SSR State Press, Tashkent (1935).

Trombe, Felix and Foex, Marc, Apparatus for plant cultivation in arid, sunny regions, (in French), Procesverbal Acad. Agric. France, 3 pages (Oct. 30, 1957).

Trombe, F. and Foex, M., Utilization of solar energy for simultaneous distillation of brackish water and air-conditioning of hothouses in arid regions, U.N. Conference on New Sources of Energy, Paper 35/S/64/Revised, Rome, 11 pages (Aug. 1961).

Tunisian Atomic Energy Commission, *Report of Activities, 1966-1967* (in French), Chapter 10, Solar Energy, pp.53-76, Solar Distillation, pp.54-64 (1967).

Tunisian Atomic Energy Commission, Brochures describing solar distillation stations at Chibou, Chekmou, and Mahdia, Tunisia (in French and Aramaic), 8 pages each (1968).

Tyagi, R.C., Low cost portable solar still, *Sunworld*, 3(6), 172 (1979).

Umarov, G.Ya. and Varadiashvile, Heat and mass transfer in convective solar distillers, *Applied Solar Energy (Geliotekhnika)*, 6(1), 30 (1970).

Umarov, G.Ya., Achilov, B.M., Djurayev, T.D. and Akhtamov, R., Water distillation by solar energy and winter 'cold', *Applied Solar Energy (Geliotekhnika)*, 8(6), 62 (1972).

Umarov, G.Ya., Achilov, B.M. and Zhuraev, T.D. Experimental investigation of heat- and mass-exchange processes in an inclined-step solar still, *Applied Solar Energy (Geliotekhnika)*, 9(3), 100 (1973).

United States Army Air Forces, *Sunshine-operated Stills and Rainfall as Sources of Water for Life Rafts*, Weather Information Branch, Publication SS-18, P.B. 38657, pp.1-17 (1943).

United States Bureau of Aeronautics, *Still, Drinking Water, Solar, Mark 1, Model 0*, NAVAER Specification M-663, January 1945, Amendment 2, August 1946.

United Nations, *Water Desalination in Developing Countries*, U.N. report, Sales No. 64, ST/ECA/82 (1964).

United Nations, Department of Economic and Social Affairs, *Water Desalination: Proposals for a Costing Procedure and Related Technical and Economic Considerations*, U.N. Publication, 56 pages (1965).

United Nations, Panel meeting on solar distillation, U.N., New York (Oct. 14-18, 1968). (Chapter outlines reported in *Solar Energy*, 12(2), 282 (Dec. 1968) and news item in 12(3), 409 (1969).

United Nations, *Solar Distillation as a Means of Meeting Small-scale Water Demands*, UN Publication ST/ECA/121 (1970).

U.S. Air Force, Instructions for obtaining drinking water with sea water distillation kit, Type LL-2, Air Material Command, Wright Field, Dayton, 6 pages (Dec. 5, 1946).

U.S. Air Force, Instructions for obtaining drinking water with distillation kit, sea water, solar type A-1, WPAFB-(A)-0-11 (April 1951).

U.S. Army Air Forces, Sunshine-operated and rainfall as sources of water for life rafts, Weather Information Branch, Publication SS-18, PB 38657, 17 pages (1943).

U.S. Dept. of Interior, *Saline Water Conversion Report for 1963*, Washington, D.C.

U.S. Bureau of Reclamation, Solar distillation plant: Schedule, general provisions, specification and drawings, Saline Water Program, San Diego, California, Specification No. DC-4902, 38 pages (1957).

U.S. Dept. of Commerce, *World Weather Records, 1951-1960*, 5 volumes (1965-1967).

U.S. Dept. of the Interior, Office of Saline Water, *Saline Water Conversion Reports* (1952-1965, 1967, 1969).

U.S. Dept. of the Interior, Office of Saline Water, A standardized procedure for estimating costs of saline water conversion, PB 161375, 19 pages (March 1956).

U.S. Dept. of the Interior, Office of Saline Water, *Proc. First International Symposium on Water Desalination*, Washington, D.C. (Oct. 3-9, 1965).

U.S. Dept. of State, Agency for International Development, Office of Engineering, *The Sun, the Sea, and Fresh Water - Solar Distillation Technology*, 30 pages (Nov. 1966).

U.S. Naval Weather Service, *World-wide Airfield Summaries*, 7 volumes (various continents) (1967-1968).

University of California, Institute of Engineering, Sea water conversion program: a technical and economic investigation of certain schemes for producing potable fresh water from sea water, *Berkeley Progress Report for the Year June 30, 1959*, Series No. 75, Issue No. 15, 28 pages (July 24, 1959).

University of California, Sea water conversion program, *Berkeley Progress Report for the Year Ending June 30, 1960*, Sea Water Conversion Laboratory, Series 75, Issue 23 (Aug. 1960).

University of California, Sea Water Conversion Laboratory, Saline water conversion research, *Progress Report*, 51 pages (1964).

University of California, Department of Engineering, Saline water research - Progress Summary, January 1, 1967-December 31, 1967, Report No. 68-1 (January 1968).

Unterberg, W., Studies of liquid film flow and evaporation with reference to saline water distillation, University of California, Los Angeles Department of Engineering, Report 61-26 (Oct. 1961).

Vaillant, J.R., Current research development concerning man's intervention in solar distillation on a meteorological scale, in order to obtain soft water from sea water in arid or tropical regions (in French), Second European Symp. on Fresh Water from the Sea, Athens (May 9-12, 1968).

Vardiashvili, A.B., Tests on a solar still, *Heliotechnology* (English trans. of *Geliotekhnika*), No. 2, 44-47 (1966).

Vardiashvili, A.B., Investigation of a solar still, *Applied Solar Energy (Geliotekhnika)*, $\underline{2}$(2), 43 (1966).

Vardiashvili, A.B. and Sharafi, A.Sh., Investigation and thermic computation of a blowing solar distiller, *Applied Solar Energy (Geliotekhnika)*, $\underline{3}$(6), 67 (1967).

Vardiashvili, A.B., Experimental investigation of optical characteristics of a solar flowing distiller, *Applied Solar Energy (Geliotekhnika)* $\underline{4}$(1), 36 (1968).

Vardiashvili, A.B., Integral heat and mass characteristics of convection solar stills, *Applied Solar Energy (Geliotekhnika)*, $\underline{7}$(4), 71 (1971).

Veinberg, B.P. and Veinberg, V.B., *Solar Stills*, Leningrad. (1933). In Russian.
Veynberg, B.P. (Title unknown), *Priroda (Nature)*, No. 2 (1930).
Veynberg, B.P. and V.B., *Solar Desalinators* (in Russian), Leningrad Municipal Literary Press (1933).
Wagner, E.G. and Lanoix, J.N., *Water Supply for Rural Areas and Small Communities*, World Health Organization, Geneva (1959).
Wang, John C., New solar energy research institute in Taiwan, *The Sun at Work*, 11-13, 17 (Second Quarter, 1961).
Ward, G.T., *Possibilities for the Utilization of Solar Energy in Underdeveloped Rural Areas*, U.N. Food and Agriculture Organization, Farm Power and Machinery Informal Working Bulletin No. 16, 116 pages (Jan. 1961).
Ward, Gerald, T., *Potentialities for Saline Water Conversion and the Provision of Power in Arid Areas*, Brace Research Inst., McGill University, Technical Report No. T-8, 13 pages (Nov. 1963).
Watanabe, Koichi, Kashima, Koji and Matsui, Koichi, Experimental and analytical study of performance for single roofed solar stills, *Proc. ISES*, New Delhi, India, $\underline{3}$, Series 51, Abstract No. 1028, 1618 (1978).
Weihe Henrik, Fresh water from sea water: Distilling by solar energy, *Solar Energy*, 13(4), 439-444 (1972).
Wheeler, N.W. and Evans, W.W., *Evaporating and Distilling by Solar Heat*, 102, 633, 8 Cl, May 3, 1870.
Whillier, Austin, Prospects for the engineering utilization of solar energy in South Africa, *South African mechanical Engineer*, Kelvin Publications, Johannesburg, 59-90 (Oct. 1956).
Whillier, Austin, Plastic covers for solar collectors, *Solar Energy*, $\underline{7}$(3), 148-151 (July/Sept. 1963).
Whillier, A. and Ward, G.T., *How to Make a Solar Still*, Brace Research Institute, (Do it yourself leaflets), January 1965, Revised February 1973.
Whinnen, A.M., Solar distillation, Honours Thesis, University of Western Australia (1964).
Wilson, A.W., Economic aspects of decision making on water use in semi-arid and arid lands, UNESCO, *Arid Zone Research, Geography of Saline Water Research*, 28 (1966).
Wilson, B.W., The desalting of bore water by solar distillation, CSIRO, Div. Indl. Chem., Melbourne, Australia, Serial No. 71, 16 pages (July 1954).
Wilson, B W., Solar distillation in Australia, *Trans., Conference on the Use of Solar Energy*, Tucson, Arizona, $\underline{3}$, 154-158 (1955).
Wilson, B.W., Solar distillation research and its application in Australia, *Proc. Symposium on Saline Water Conversion*, National Academy of Sciences, National Research Council, Washington, D.C., Publication No. 568, 123-130 (Nov. 4-6, 1957).
Wilson, B.W., Personal communication, CSIRO, Division of Chemical Engineering, Clayton, Victoria, Australia.
Xogler, X. and Kubler, M., Sea water desalination by solar heated heat pipes, *The Sun in the Service of Mankind*, UNESCO House, Paris, pp.2-6 (1973).
Zandi, I., Improved solar still process for desalting sea and brackish water, Ph.D. Thesis, School of Civil Engineering, Georgia Institute of Technology (Oct. 1959).
Zarem, A.M. and Erway, Duane, D. (editors), *Introduction to the Utilization of Solar Energy*, McGraw-Hill, N.Y. (1963), Chapter 12, the Distillation of Sea Water and other Low-Temperature Applications of Solar Energy (Everett D. Howe), Section 12.4, Solar Distillation, pp.301-307.
Zongnan, Li and Qing, Wue, Solar stills, *Acta Energiae Solaris Sinica*, $\underline{1}$(1), 29(1980), or *ASSET*, $\underline{2}$(8), 15 (Sept. 1980).

PATENTS

Abbot, C.G., Solar distilling apparatus, U.S. Patent No. 2,141,330 (Dec. 27, 1938).
Adamec, O.T., U.S. Patent No. 3,192,133.
Agnew, E.A., U.S. Patent No. 2,636,129, French Patent No. 1.077.713, German Patent No. 955.917, Swiss Patent No. 328.023.
Aven, P.S., U.S. Patent No. 3,077,190.
Altenkirch, E. and Behringer, H., Apparatus for evaporating sea water by the heat of the sun, Brit. Patent No. 454,558 (Mar. 27, 1935).
Avery, E.N., U.S. Patent No. 3,509,716.
Barnes, W.S., Solar water still, U.S. Patent No. 2,383,234 (Aug. 21, 1945).
Beard, K.D., U.S. Patent No. 3,193,473.
Bimpson, H.S. and Palmer, E.J., Solar salt water distilling apparatus, U.S. Patent No. 2,424,142 (July 15, 1947).
Bjorksten, J., Water purifier, U.S. Patent No. 2,848,389 (Aug. 19, 1958).
Bjorksten, J., U.S. Patent No. 2,788,316.
Bohmfalk, B.H., Distilling apparatus, U.S. Patent No. 2,332,294 (Oct. 19, 1943).
Brosius, A.M., Solar still, U.S. Patent No. 983,434 (Feb. 1, 1911).
Buckley, J.L., U.S. Patent No. 2,843,536, British Patent No. 743,539.
CSIRO, An improved diffusion still, Australian Patent No. 65,270/60 (Oct. 1960).
Delano, W.R.P., Collapsible distillation apparatus, U.S. Patent No. 2,398,291 (Apr. 9, 1946).
Delano, W.R.P., Solar distilling apparatus, U.S. Patent No. 2,398,292 (Apr. 9, 1946).
Delano, W.R.P., Process and apparatus for distilling liquids, U.S. Patent No. 2,402,737 (June 25, 1946).
Delano, W.R.P. and Meissner, W.E., Solar distillation apparatus, U.S. Patent No. 2,405,118 (Aug. 6, 1946).
Delano, W.R.P., Apparatus for solar distillation, U.S. Patent No. 2,405,977 (Aug. 13, 1946).
Delano, W.R.P., Solar still with nonfogging window, U.S. Patent No. 2,413,101 (Dec. 24, 1946).
Delano, W.R.P., Inflatable solar still, U.S. Patent No. 2,427,262 (Sept. 9, 1947).
Dooley, G.W., Means for purifying water, U.S. Patent No. 1,812,516 (June 30, 1931).
Dornig, M. and Belloni, A., Equipment to produce distilled water or the like through solar heat (in Italian), Italian Patent No. 448,833 (May 27, 1949).
Edlin, F.E., Air supported solar still, E.I. du Pont de Nemours, U.S. Patent No. 3,174,915 (Mar. 23, 1965).
Edlin, F.E., Air-supported solar still, E.I. du Pont de Nemours, U.S. Patent No. 3,174,915 (Mar. 23, 1965). Filed July 1962, Serial No. 211,942.
E.I. Du Pont de Nemours & Co., British Patent No. 905,760, French Patent No. 1.261.062, U.S. Patent No. pa 823.237/59.
Englisch, O., Improvement in or relating to evaporators, British Patent No. 344,125 (1931).
Gandillon, P.A. and Garchey, L.A., Evaporating water from the surface exposure to the Sun's rays, British Patent No. 229,211 (Sept. 15, 1924).
Ginnings, D.C., Multiple-effect solar still, U.S. Patent No. 2,445,350 (July 20, 1948).
Gordon, A., Applying solar heat to the evaporation of fluids, generation of steam, etc., British Patent No. 1044 (May 1856).
Graham, F.A., Distilling apparatus, U.S. Patent No. 1,302,363 (Apr. 29, 1919).
Grosse, A.V., Solar still, U.S. Patent No. 4,141,798 (Feb. 27, 1979), File date: Jan. 30, 1978.

Halacy, D.S., U.S. Patent No. 3,337,418.
Hay, Harold, R., Process and apparatus for solar distillation, U.S. Patent No. 3,314,862 (Apr. 18, 1967).
Hermansen, N.T., U.S. Patent No. 3,338,797.
Kain, S.C., Distillation device, U.S. Patent No. 2,342,201 (Feb. 22, 1944).
Kenk, R., U.S. Patent No. 3,359,183.
Kimmerle, H.J., U.S. Patent No. 3,232,846.
Lighter, S., U.S. Patent No. 2,820,744.
Malek, J.M., Solar evaporator, U.S. Patent No. 3,167,488 (Jan. 26, 1965).
Manly, A.H., U.S. Patent No. 2,902,028.
Massie, L.E., Solar still with replaceable solar absorbing liner and weight controlled feed inlet, U.S. Patent No. 3,880,719 (Apr. 1975).
Merz, F., System and apparatus for the utilization of natural thermal energy, British Patent No. 152,753 (Oct. 1920).
Miller, W.H., Jr., Inflatable floating solar still with capillary feed, U.S. Patent No. 2,412,466 (Dec. 10, 1946).
Miller, W.H., Improvement in or relating to solar distillation apparatus, British Patent No. 594,131 (Nov. 1947).
Mount, W.W., British Patent No. 743, 539, U.S. Patent No. 2,843,536.
Muller, J.G., U.S. Patent No. 3,138,546.
Schenk, T.C., Device for producing potable water from sea water, U.S. Patent No. 2,342,062 (Feb. 15, 1944).
Shachar, S., U.S. Patent No. 3,372,691.
Smith, Robert W., Glekas, Louis P., Riley, John E. and Swingle, Robert L., Compact solar still, U.S. Patent No. 3,397,117 (Aug. 13, 1968).
Snyder, R.E., Solar heated vacuum still, U.S. Patent No. 2,490,659 (Dec. 6, 1949).
Societe Industrielle et Commerciale des Salins du Mid, French Patent No. 1.581.816.
Telkes, M., Collapsible solar still with water vapor permeable membrane, Melpar, Inc., U.S. Patent No. 3,415,719 (Dec. 10, 1968).
Ushakoff, A.E., Inflatable solar still, U.S. Patent No. 2,455,834 (Dec. 7, 1948).
Ushakoff, A.E., Inflatable solar still, U.S. Patent No. 2,455,835 (Dec. 7, 1948).
Wheeler, N.W. and Evans, W.W., Evaporating and distilling with solar heat, U.S. Patent No. 102,633 (1870).
Ziehm, K.F., Jr., Desalination apparatus, U.S. Patent No. 4,077,849 (Mar. 7, 1978). File date: Nov. 3, 1975.
Ziem, T., Equipment and apparatus for distillation of water and other liquids with solar heat, British Patent No. 12,402 (Aug. 28, 1888). Also German Patent No. 47,446 (June 22, 1889).

AUTHOR INDEX

Abbot, C.G. 4
Achilov, B.M. 53, 74, 75, 77
Ahmedzadeh, J. 94, 95
Akhtamov, R.V. 5, 74, 75
Akinsete, V.A. 54, 57
Appleyard, J. 64, 65, 66
Araujo, S.R.D. 67

Baibutaev, K.B. 56, 58
Bairamov, R. 111, 114
Bartali 5, 61, 62
Baum, V.A. 4, 11, 23, 53, 54, 111
Bloemer, J.W. 6, 32, 36, 51, 56, 113, 114
Boelter, L.M.K. 10
Bowen, I.S. 13
Brusewitz 76, 78, 81, 83

Cooper, P.I. 4, 8, 14, 18, 20, 22, 23, 31, 32, 36, 37, 54, 56, 57, 58, 64, 65, 66

Daniels, F. 51
Datta, R.L. 115
Della Porta 2, 3
Delyannis, A.A. 19, 21, 97
Delyannis, E. 19, 21
Donald, Q.K. 63
Duffie, J.A. 15, 25, 29, 34, 87
Dunkle, R.V. 11, 13, 46, 69, 82, 85, 86, 87, 104
Duru, C.U. 57

Edlin, F.E. 98

Falvey, H.T. 93, 94
Frick, B. 4, 23
Frick, G. 5, 79, 80

Garg, H.P. 33, 56
Garg, S.K. 116, 117
Goghari, H.D. 125
Gomkale, S.D. 115, 116, 117, 122, 125
Grune, W.N. 39
Harding, J. 3, 4
Henrik, W. 59
Hirschmann, J.R. 4, 23
Howe, E.D. 5, 40, 41, 51, 52, 72, 73, 74
Hsieh, C.K. 33

Jakob, M. 9

Kausch, O. 4
Keller, J.D. 42
Kobayashi, M. 94

Lawand, T.A. 98
Lewis, W.K. 12
Lobo, P.C. 62, 63, 67
Lof, G.O.G. 7, 19, 50, 54, 79, 80, 97, 112, 113, 115

MacLeod, L.H. 80, 81
Malik, M.A.S. 5, 40, 44, 45, 67, 68
Mann, H.S. 33, 56
McAdams, W.C. 27, 29
McCracken 80, 81
Menguy, G. 5, 92, 93
Menozzi, G. 2
Morse, R.N. 37, 54, 56
Mouchot, A. 2, 3
Moustafa, S.M.A. 5, 76, 78, 81, 83, 87

Natu, G.L. 125
Nayak, J.K. 4, 20, 23, 28, 30, 31, 32
Nebbia, G. 2
Norov, E.Zh. 5, 91, 92

Oltra, F. 5, 60

Pasteur, F. 4

Qasim, S.R. 100, 101, 103

Rajvanshi, A.K. 33
Rao, K.S. 125

Roefler, S.K. 4, 23
Read, W.R.W. 37, 54, 56

Schmidt, E. 29, 86
Selcuk, M.K. 5, 104, 105
Sharpley, B.F. 10
Sodha, M.S. 4, 5, 24, 35, 36, 37, 39, 67, 68, 70, 71, 81, 84, 86, 88, 89, 104, 106
Soliman, H.S. 5, 54, 96
Sommerfeld, J.V. 5, 79, 80

Talbert, S.G. 59
Telkes, M. 4, 79, 90, 91, 92
Threlkeld, J.L. 27, 33
Tinaut, D. 102, 103, 104
Tleimat, B.W. 40, 41, 51, 52, 59, 72, 73, 74
Todd, C.J. 93, 94
Tran, V.V. 40, 44, 45

Umarov, G.Ya. 5, 92
United Nations 54, 55, 98, 99, 110, 111, 118, 119

Wilson, Carlos 3
Wong, H.Y. 8

SUBJECT INDEX

Air inflated plastic still 4, 97
Angle of incidence, effect on radiation parameters 22

Basin type
  solar still 4
  stepped solar still 76, 78
Bottom loss coefficient 46

Chimney type solar still 61
Concentric tube solar still 93, 94
Convection 8, 9, 14
  forced 9, 121
  free 9
Cost comparison
  for different techniques of desalination 117
  for solar distillation 115

Daily distillate output 5
  ground type still 95
  single basin still 29
  single basin still using dye 35
  solar collector-basin type still 96
  still on roof system 108
  tubular solar still 40
Desalination 2
Diffusion solar still 60
Distilling 3
Double basin solar still 67-71, 121
  analysis 68

construction 68
experiment 70

Economic analysis of solar distillation 112
  Indian experience 115
  Russian experience 114
Economics of solar still plant at Awania 125
Effect of angle of incidence on radiation parameters 22
Effect
  of ambient temperature 54, 55
  of charcoal pieces 57
  of cover inclination 56
  of double glass cover 56
  of dye 33-36, 109
  of floating mat 98
  of formation of algae and mineral layers 58
  of salt concentration 56
  of solar radiation and loss coefficient 54, 55
  of thermal capacity 56
  of wind velocity 54
Efficiency 5
  basin type solar still 5
  ideal still 23
  simple multiple wick solar still 86
Electrodialysis for distillation 2, 110
Energy
  flow diagram 7
  incoming to the still 6
  outgoing from the still 6
Energy balance

for conventional still, steady state 40
for double basin still 68
for ground still 23
for ideal still 22
for single basin still 37
for water moving in the basin 46
Equivalent mass transfer coefficient 12
Equivalent temperature difference 10, 11
Evaporation 11-13
Evaporative heat flux
for double basin still 70
for single basin still 39
for single basin still (ground still) 27
for single basin still (mounted still) 28
for still-on-roof system 108
External convection coefficient 15
Extruded plastic still 99

Film covered solar still 5, 91, 92
Flash distillation 110
Fourier's equation 15
Fresh water 1

Greenhouses 2
Greenhouse type solar still 4
Ground still 19, 20, 23-28, 31, 32

Heat and mass transfer phenomenon 8-17, 32, 121
Heat balance see Energy balance
Heated head solar still 61, 62
Heat transfer
convective 11, 14
evaporative 13, 14
radiative 13, 14
Heat transfer coefficient
convective 12, 24
evaporative 24
outside the still 25
radiative 24
Hourly distillate yield
of single basin solar still 39
of single basin solar still under nocturnal production 49

Ideal solar still 22
analysis 22
efficiency 23
Inclined solar still 72-89

Inclined-stepped solar still 5, 72-74

Life raft type solar still 4, 90, 91

Mass transfer
coefficient 12
rate 12
Mounted still 19, 20, 28, 30, 31
Multi-basin system 76
Multi-effect evaporation 2, 4
Multiple effect solar still 59-71
design principle and construction 62
performance 63
Multiple ledge tilted still 80, 81
Multiple wick solar still 5, 81-89, 121
advantages 88
analysis 85
efficiency 86
experiment 82
Multi-stage flash evaporation 2

Nocturnal production of solar still 39
rigorous analytical model 46
simplified mathematical model 42

Parametric studies 50-58
Performance
of simple multi-effect basin type solar still 63
of solar still plant at Awania 125
Periodic
boundary conditions 16
heat transfer in conducting media 15-17
theory of double basin still 67
theory of solar still 23, 33
Photosynthetic activity, measurement of 102
Plastic covered circular still 52
Plastic covers, an experience 50-53
Porous wick type inclined solar still 80
Potable water see Fresh water

Radiation 13, 14
Regenerative inclined stepped solar still 74, 75
Reverse osmosis for distillation 2, 110
Roof type solar still 4

Saturation vapor pressure 24, 29, 34

Subject Index

Sky temperature 15
Special Grashof number 10, 11
Specific humidity 11
Single basin solar still 4, 18-58, 72, 121
  effect of dye 33-36
  experiment 28
  nocturnal production 39
  periodic theory 23
  solar radiation balance 20
  transient analysis 37
Solar collector-basin type still 96
Solar distillation 2, 100, 115, 118, 121
  basic principles 5-7
  economic aspects 110-120
  historical review 2-5
  major plants 19, 21
  versus conventional process 118
Solar earth water stills 94-95
Solar radiation balance, single basin solar still 20
Solar still 3
  categorization 5
  chimney type 61
  concentric tube 93
  deep basin 7, 73
  diffusion type 60
  double basin 67-71, 121
  expandable 5
  film covered 5, 91, 92
  greenhouse type 4
  heated head 61, 62
  inclined stepped 5, 72
  indirectly heated 5
  life raft 4, 90, 91
  multiple wick type 5, 81-89, 121
  plant at Awania 122-126
  regenerative inclined stepped 74
  roof type 4
  single basin 4, 18-58, 72, 121
  stepped inclined 75
  tilted tray 4, 5, 52, 72
  tilted wick type 5, 79
  tubular 4, 40, 41
  V-covered 4
  wiping spherical 5, 92
  with reflectors 3, 4, 98
Solar still - greenhouse combination 5, 100-109
  construction 100
  materials for 101
  Taxas experience 100
Still construction, use of different materials 53
Still-on-roof system 104-109
  analysis 106
  distillate output 108
  evaporative heat flux 108

Thermal convection 8
Thermal losses *see* Energy losses
Thin film distillation 2
Three effect multiple solar still 64-66
  construction 64
Tilted single wick still 5, 79
Tilted tray solar still 72, 73
Transient analysis
  of solar still 37-39
  of double basin still 71
Tubular solar still 4, 40, 41

Vapour compression process for distillation 110, 112
V-covered solar still 4

Wick type collector - evaporator still 81, 83
Wiping spherical still 5, 92